2nd International Conference on Materials Science and Engineering (IC-MSE 2021)

Held online due to COVID-19

IOP Conference Series: Materials Science and Engineering
Volume 1133

Suzhou, China
17 - 19 March 2021

ISBN: 978-1-7138-3411-3
ISSN: 1757-8981

Printed from e-media with permission by:

Curran Associates, Inc.
57 Morehouse Lane
Red Hook, NY 12571

Some format issues inherent in the e-media version may also appear in this print version.

Institute of Physics
Dirac House, Temple Back
Bristol BS1 6BE UK

Phone: 44 1 17 929 7481
Fax: 44 1 17 920 0979

techtracking@iop.org

Additional copies of this publication are available from:

Curran Associates, Inc.
57 Morehouse Lane
Red Hook, NY 12571 USA
Phone: 845-758-0400
Fax: 845-758-2633
Email: curran@proceedings.com
Web: www.proceedings.com

TABLE OF CONTENTS

PREFACE

2021 2nd International Conference on Materials Science and Engineering (IC-MSE 2021) was successfully held at Suzhou, China during March 17-19, 2021. Due to the outbreak of COVID-19, the organizing committee will follow the guidelines of WHO (World Health Organization), and local government agencies on the current situation. The IC-MSE 2021 was hold virtually. All the keynote sessions and technical session of IC-MSE 2021 were successfully conducted through hybrid platform (Online and offline).

The objective of IC-MSE 2021 Conference is to provide a venue for researchers and practitioners in related fields from academia, industry and government to meet in a forum where the latest research results are presented and prospect for future developments on "materials science and engineering". Professors from Japan and Portugal are invited to deliver keynote speeches regarding latest information in their respective expertise areas, where the speakers discussed various issues of materials science. There were two oral sessions and one poster session. Based on the technical, quality and presentation, best papers were selected and awarded during each technical session.

This proceeding presents a selection from papers submitted to the conference from universities, research institutes and industries. All of the papers were subjected to peer-review by conference committee members and international reviewers. The papers selected depended on their quality and their relevancy to the conference. The volume tends to present to the readers the recent advances in the field of materials science and engineering and various related areas, such as biomaterials and nanomaterials, material chemistry, materials processing engineering and structural material, etc.

We express our sincere gratitude to all people who contributed to the review process and did all their best to enhance the scientific merit and quality of the Proceedings. We are also grateful to the authors for their contributions, to the speakers and participants of the IC-MSE 2021. We hope to welcome them along with new participants to the next event IC-MSE 2022.

Prof. Masaru Tanaka
Institute for Materials Chemistry and Engineering, Kyushu University, Japan
March. 21, 2021

IOP Conf. Series: Materials Science and Engineering 1133 (2021) 011001 doi:10.1088/1757-899X/1133/1/011001

Conference Committee

Conference Chair

Prof. Nuno Dinis Corticos, Universidade de Lisboa, Portugal

Program Chair

Prof. Masaru Tanaka, Institute for Materials Chemistry and Engineering, Kyushu University, Japan

Assoc. Prof. Wong Wah Sang, Department of Architecture, University of Hong Kong

Technical Committee

Prof. Yoshiki Katayama, Department of Chemistry and Biochemistry, Faculty of Engineering, Kyushu University, Japan

Prof. Yukiko Ozaki, Kyushu University, Japan

Prof. CHUNG Tai-Shung, Neal, Department of Chemical and Biomolecular Engineering, National University of Singapore, Singapore

Prof. Cheng Siong Chin, Newcastle University in Singapore, Singapore

Prof. Pilar Lopez Cornejo, Unoversity of Seville, Spain

Prof. Ing. Milena Pavlikova, Department of Materials Engineering and Chemistry of the Faculty of Civil Engineering, Czech Technical University at Prague

Prof. Eric James Strauss, Michigan State University, USA

Prof. Yan Yunjun, Huazhong University of Science and Technology, China

Prof. Muhammad Bashir Ibrahim, BAYERO UNIVERSITY, KANO - NIGERIA

Prof. MUHAMMAD BASHIR IBRAHIM, BAYERO UNIVERSITY, KANO - NIGERIA

Prof. (Neal) Tai-Shung Chung, Department of Chemical and Biomolecular Engineering, National University of Singapore, Singapore

Prof. R. C. Gupta, Consultant RNB Global University Bikaner India

Prof. Snehal Agnihotri, Commerce and Science College, India

Prof. B C Meikap, Indian Institute of Technology Kharagpur, India

Prof. Sang Hyuk Im, Korea University, South Korea

Prof. Mitsutoshi Jikei, Department of Materials Science, Akita University, Japan

Prof. Vasile Lavric, University Politehnica of Bucharest, Romania

Prof. Hartmut Hinz, Frankfurt University of Applied Sciences, Germany

Prof. John Ryan Dizon, Department of Industrial Engineering, College of Engineering and Architecture, Bataan Peninsula State University, Philippines

IOP Conf. Series: Materials Science and Engineering 1133 (2021) 011001 doi:10.1088/1757-899X/1133/1/011001

University of Malaya
Senior Lecturer. Shaifulazuar Rozali, Mechanical Engineering Department, Faculty of Engineering, University Malaya
Senior Lecturer. Hashem Salarzadeh Jenatabadi, Department of Science & Technology Studies, Faculty of Science, University of Malaya
Senior Lecturer. Teoh Wen Hui, Chemical Engineering Department, Faculty of Engineering, University of Malaya
Senior Lecturer. Nazatul Liana Binti Sukiman, Department of Mechanical Engineering, Faculty of Engineering, University of Malaya
Dr. Aolei Gao, University of Chinese Academy of Sciences, China
Dr. Lei Du, Harbin Institute of Technology, China
Dr. Han-Yu Hsueh, National Chung Hsing University, Taiwan, China
Dr. Han-Yu Hsueh, Department of Material Science and Engineering, National Chung Hsing University, Taiwan, China
Dr. Yue Zhang, The University of Texas at Austin, USA
Dr. Al-fadhl Yahya Khaled Al-khaled, University of Kentucky, USA
Dr. Ksenia ILLKOVÁ, Czech Academy of Science, Institute of Plasma Physics, Department of Materials Engineering, Czech Republic
Dr. Mohamed Infas Haja Mohideen, Charles University, Czech Republic
Dr. Wenfu Situ, Technical University of Denmark, Denmark
Dr. Khoirina Dwi Nugrahaningtyas, Sebelas Maret University, Indonesia
Dr. Eng. Kusno Kamil, MEng.Mngt, Universitas Muslim Indonesia
Dr. Rafal Chodun, Faculty of Materials Science, Warsaw University of Technology, Poland
Dr. Muhammad Ahsan, National University of Sciences and Technology, Pakistan
Dr. Bruno Manuel Santos Castro Martins, CEGOT - Centre of Studies in Geography and Spatial Planning - University of Coimbra, Portugal
Dr. Sarintorn Limpanart, Metallurgy and Materials Science Research Institute, Chulalongkorn University, Thailand
Dr. Nithiwach Nawaukkaratharnant, Metallurgy and Materials Science Research Institute, Chulalongkorn University, Thailand
Dr. Manunya Okhawilai, Energy Research Institute, Chulalongkorn University, Thailand
Dr. Chatree Wattanasin, Energy Research Institute, Chulalongkorn University, Thailand
Dr. Piangjai Peerakiatkhajohn, Faculty of Environment and Resource Studies, Mahidol University, Thailand
Dr. Saowanee Wijitkosum, Environmental Research Institute, Chulalongkorn University, Thailand
Dr. Sira Srinives, Department of Chemical Engineering, Faculty of Engineering, Mahidol University, Thailand
Dr. Sarintorn Limpanart, Metallurgy and Materials Science Research Institute, Chulalongkorn University, Thailand
Dr. Manunya Okhawilai, Metallurgy and Materials Science Research Institute, Chulalongkorn University, Thailand
Dr. Bussarin Ksapabutr, Department of Materials Science and Engineering, Faculty of Engineering and Industrial Technology, Silpakorn University, Thailand
Lecturer. Surawith Tunchaiyaphum, Southeast Asia University, Thailand

IOP Conf. Series: Materials Science and Engineering 1133 (2021) 011001 doi:10.1088/1757-899X/1133/1/011001

Dr. Supawat Vivanpatarakij, Energy Research Institute, Chulalongkorn University, Thailand
Dr. Rasamee Chaisuksant, Department of Chemistry, Faculty of Science, Silpakorn University, Thailand
Dr. Nichanun (Natt) Sirasunthorn, Development and Promotion of Science and Technology Talents Project (DPST), Silpakorn university, Thailand
Dr. Radchada Buntem, Department of Chemistry, Faculty of Science, Silpakorn University, Thailand
Dr. Mukta Bansal, NTU, Singapore
Dr. Hoang Thi Hong Cam, University of Science and Technology of Hanoi, Vietnam
Dr. Nguyen Van Trang, Ho Chi Minh City University of Technology and Education, Vietnam
Dr. LE PHUONG THU, UNIVERSITY OF SCIENCE AND TECHNOLOGY OF HANOI (USTH), Vietnam
Dr. NGUYEN Van Phuong, University of Science and Technology of Hanoi, Vietnam
Dr. Bui Van Hoi, University of Science and Technology of Hanoi, Vietnam
Dr. Chaher Zid, Universiti Tun Hussein Onn Malaysia, Malaysia
Dr. AL-FADHL YAHYA KHALED AL-KHALED, Universiti Putra Malaysia, Malaysia
Dr. Elsayed Talaat Mohamed Helmy, National Institute of Oceanography and Fisheries (NIOF), Egypt
Dr. Nabil Al-Zaqri, Chemistry Department, College of Science, King Saud University, Saudi Arabia
Dr. Fengqiao Zhang, Delft University of Technology, The Netherlands

Peer review declaration

All papers published in this volume of *IOP Conference Series: Materials Science and Engineering* has been peer reviewed through processes administered by the Editors. Reviews were conducted by experts referees to the professional and scientific standards expected of a proceedings journal published by IOP Publishing.:

- **Type of peer review:** Double-blind (Conference reviewers and iThenticate system)
- **Conference submission management system:** Conference Management Toolkit (CMT)
- **Number of submissions received:** 12 full papers and 12 abstracts
- **Number of submissions sent for review:** 12 full papers
- **Number of submissions accepted:** 7 full papers
- **Acceptance Rate (Number of Submissions Accepted / Number of Submissions Received X 100):** 58%
- **Average number of reviews per paper:** 3
- **Total number of reviewers involved:** 58
- **Any additional info on review process:** The papers were evaluated based on the technical criteria, quality criteria and presentation criteria. The referees argued detailed their reviews and provided an overall evaluation (strong accept, accept, weak accept, weak reject, reject, strong reject), specifying their confidence (expert, high, medium, low, none) regarding qualifications to review the allocated paper.
- **Contact person for queries:** Ms. Alice Liu; alice@isacri.org

IOP Conf. Series: Materials Science and Engineering 1133 (2021) 012001 doi:10.1088/1757-899X/1133/1/012001

Progress in the study of heteropolyacid-based-nanocomposite photocatalytic materials

[1]Li Pengxi, Zhou Peng, Guo Zemei and Yang Jinging

Environment and Quality Test Department, Chongqing Chemical Industry Vocational College, 401228, Chongqing, China
[1]lipengxi616@163.com;

Abstract. As a kind of mild green photocatalysts, Heteropoly acid have a good prospect in photocatalysis due to its high activity, good selectivity and non-toxicityand possess photochemical characteristics similar to those of the semiconductor photocatalysts. Heteropoly acid (HPA), which are very soluble in water and polar solvents, these compounds are activated only by UV light, which restricts the practical application of HPA. Dispersing HPA onto nanoparticle supports with high surface area is useful to increase their specific surface area and hence (photo)catalytic activity. In this paper, the application of HPA-nanoparticles composites such as HPA/TiO2 nanoparticles composites, HPA/SiO2 nanoparticles composites, HPA/ZrO2 nanoparticles composites and so on, in photocatalytic degradation of printing and dyeing wastewater, and the future research direction is prospected.

1. Introduction

With the development of the printing and dyeing industry, a large amount of printing and dyeing wastewater seriously threatens the ecological, environmental systems and human health, and it is urgent to develop a simple and efficient treatment technology for printing and dyeing wastewater [1], [2]. Semiconductor photocatalysis, using sunlight as the light source and inexpensive semiconductor materials as the medium, has the advantages of mild reaction conditions, low cost and no pollution, which is an ideal treatment technology for printing and dyeing wastewater [3]. In recent years, researchers have developed a wide variety of semiconductor photocatalytic materials, such as metal oxides (such as TiO_2, ZnO, SnO_2), metal sulphides (such as CdS, MoS_2, WS_2), inorganic non metals (such as graphene, graphene oxide, g-C_3N_4), etc. [4]-[6]. Traditional semiconductor photocatalytic materials still have some drawbacks in use: (1) low visible light utilization, (2) high photogenerated electron-hole complexation rate, (3) low quantum efficiency, which makes it difficult to treat large amounts of printing and dyeing wastewater. Therefore, the solution of the above scientific problems is currently a hot research topic in the field of semiconductor photocatalysis.

Heteropolyacids (HPA), a semiconductor-like material, have properties similar to those of semiconductors [7]. HPA, as a class of excellent electron acceptors, can effectively promote photogenerated electron-hole separation and fast electron transport, with the advantages of good selectivity, high catalytic activity and no pollution [8]. However, the light-responsive absorption region of HPA is mainly concentrated in the UV region, with low utilization of visible light sources; HPA also has the disadvantages of low porosity, lack of active sites on the surface, low specific surface area (5-8 m^2/g), and easy solubility in polar solvents causing secondary pollution, which limits the further application of HPA [9]. To overcome this drawback, HPA-based nanocomposites are often prepared by loading HPA onto nanomaterials using methods such as impregnation, adsorption and hybridization [10]. HPA-based nanocomposites have been widely used in electrocatalysis, photocatalysis and biocatalysis because of their properties such as high charge density, diverse size and multifunctional structure, as well as the optical and mechanical properties of nanoparticles [11], [12]. In this paper, we will focus on the application of HPA-based nanocomposites in photocatalytic degradation of printing and dyeing wastewater, and look into future research directions.

2. HPA/TiO_2 nanocomposites

Titanium dioxide (TiO_2) nanoparticles are widely used for their non-toxicity, photostability, unique photochemical properties and high chemical stability [13]. However, the photocatalytic efficiency of TiO_2 is limited due to the rapid complexation of light-induced electron-hole pairs. Therefore, the use of different electron acceptors to retard the electron-hole complexation reaction can help to improve the photocatalytic efficiency [13]. As mentioned above, HPA/TiO_2 nanocomposites, which can increase the specific surface area, increase the active sites and thus improve the photocatalytic activity, have a synergistic enhancement effect and are gaining increasing attention.

Jin Hongxiao et al [14] prepared several XW_{12}/TiO_2 (X = P, Si, Ge) nanocomposites by the sol-gel method, and the visible light degradation of the textile dye X-3B revealed that the photocatalytic activity of XW_{12}/TiO_2 nanocomposites was significantly higher than that of HPA and TiO_2 alone. different heteropolyacids had a significant effect on the photocatalytic activity of XW_{12}/TiO_2 nanocomposites. The order of photocatalytic activity of HPA-based nanocomposites was as follows: phosphotungstic acid (PW_{12})/TiO_2>silicotungstic acid (SiW_{12})/TiO_2>germanium tungstic acid (GeW_{12})/TiO_2. Shahram et al [15] prepared a $Na_5[PV_2Mo_{10}O_{40}]/TiO_2$ nanocomposite. composite. Photocatalytic degradation experiments showed that $Na_5[PV_2Mo_{10}O_{40}]/TiO_2$ could photocatalyze the degradation of nine organic dyes such as bromothymol blue and congo red. The degradation of methyl orange, Congo red and methylene blue by the nanocomposite $Na_5[PV_2Mo_{10}O_{40}]/TiO_2$ was above 90% under ultrasound-assisted visible light irradiation for 20 min. Yang Changjun et al [16] prepared a novel polyvinyl chloride (PVC)/ PW_{12}/TiO_2 nanocomposite film and photodegradation behavior was investigated. The results of rhodamine B photocatalytic degradation showed that the photocatalytic activity of PVC/ PW_{12}/TiO_2 with a HPA loading of 2 wt. % was 1.5 times higher than that of PVC/TiO_2. Shi Huixian et al [17] used a sol-gel method to prepare nanoscale TiO2 photocatalysts co-doped with rare-earth ions (La^{3+}, Ce^{3+}) and HPA Ce/ La-PW_{12}/TiO_2 nanocomposites. The photocatalytic degradation of methylene blue was found to be 98% and 96% for Ce-PW_{12}/TiO_2 and La-PW_{12}/TiO_2, respectively, with significantly higher photocatalytic degradation activities than Ce-TiO_2, La-TiO_2, PW_{12}/TiO_2 and PW_{12}/TiO_2 for 100 min of light exposure. The photocatalytic mechanism study showed that there is a synergistic effect between rare earth ions and heteropolyacids as well as TiO_2, which can increase the light adsorption and inhibit the compounding of photogenerated electron-hole pairs. Therefore, improving the synergistic effect of the multi-component is one of the ideal methods to overcome the disadvantages of low quantum efficiency and high electron-hole complexation rate of photocatalytic materials. Various HPA/TiO2 nanocomposites for organic dye degradation are listed in Table 1.

Table 1. Recent outstanding results of the HPA/TiO_2 nanocomposites for organic dye degradation

Photocatalyst	Synthetic method	Contaminants	Degradation efficiency	Reference
PW_{12}/TiO_2	sol–gel method	X-3B	~89% in 50min	[14]
$Na_5[PV_2Mo_{10}O_{40}]/TiO_2$	Solvothermal	bromothymol blue	~90% in 30min	[15]
		Congo Red	~90% in 30min	
		Rhodamine B	~90% in 30min	
PVC/ PW_{12}/TiO_2	Solvothermal	Rhodamine B	~18% in 60min	[16]
Ce-PW_{12}/TiO_2	Sol-gel method	Rhodamine B	98% in 100min	[17]

Compared with HPA and TiO_2 monomers, the photocatalytic activity of HPA/TiO_2 nanocomposites has been greatly improved, but there are still drawbacks such as not easy separation and poor reusability in practical applications. In the future, the preparation of composite photocatalytic materials with high photocatalytic activity, good selectivity and easy separation and recycling will be an important direction for research.

3. HPA/SiO_2 nanocomposites

Although silicon dioxide (SiO_2) has no obvious photocatalytic activity, the encapsulation of photocatalytically active materials in mesoporous or microporous SiO_2 matrices is currently a hot topic of research in the field of photocatalysis. Furthermore, due to the unique physicochemical properties and unusual internal structure of the SiO_2 matrix, it is easy to separate and recover from the reaction environment while enhancing the catalytic activity of the photoactive material [18], [19]. Therefore, HPA/SiO_2 nanocomposites have been used as multiphase photocatalysts for various organic pollutants. Tao Shengyang et al [20] prepared PW_{12}/SiO_2 by grafting method and photocatalytic degradation experiments showed that PW_{12}/SiO_2 achieved 97.8% removal of basic magenta by 80 min of light. Besides, the PW_{12}/SiO_2 catalyst prepared by graded porous SiO_2 as a carrier was very stable and easily recovered from the reaction solution for reuse. Hou Yang et al [21] prepared a series of $Fe_3O_4/SiO_2/PW_{12}$ core-shell nanoparticles by the impregnation method, and the degradation experiments of rhodamine B showed that the UV light irradiation for 60 min, the ratio of PW_{12} 0.4% of $Fe_3O_4/SiO_2/PW_{12}$ showed the best photocatalytic activity (99.3%). The separation of rhodamine B solution from $Fe_3O_4/SiO_2/PW_{12}$ photocatalytic material was achieved in 30 s under the action of an external magnetic field, which provides a new idea for the construction of a new HPA-type photocatalyst that can be separated rapidly.

Compared to HPA homogeneous catalysts, HPA/SiO_2 nanocomposites possess different sized pore-like structures, have higher photocatalytic activity and are easy to recycle, but the biggest drawback is the low response to visible light. Li Hongfang et al [22] deposited (phosphotungstic acid- sulphur cordial, PW_{12}-TH)n polymers onto $Fe_3O_4@SiO_2$ magnetic microspheres by a layer-by-layer autoloading method. SiO_2 magnetic microspheres to produce $Fe_3O_4@SiO_2@$ (PW_{12}-TH)n nanocomposites. The visible light irradiation for 4 h resulted in a degradation rate of methyl orange of about 90% and a significantly higher catalytic activity than $Fe_3O_4@SiO_2$. Besides, the use of magnetic materials ensured a convenient, clean, fast and efficient separation of the composites and allowed a high photocatalytic activity even after multiple reuses. Various HPA/SiO_2 nanocomposites for organic dye degradation are listed in Table 2.

Table 2. Recent outstanding results of the HPA/SiO_2 nanocomposites for organic dye degradation

Photocatalyst	Synthetic method	Contaminants	Degradation efficiency	Reference
PW_{12}/SiO_2	grafting method	Fuchsin basic	98% in 80min	[20]
$Fe_3O_4/SiO_2/PW_{12}$	impregnation method	Rhodamine B	99% in 60min	[21]
$Fe_3O_4@SiO_2@(PW_{12}$-TH)n	impregnation method	methyl orange	90% in 240min	[22]

SiO_2 itself does not have photocatalytic activity, but as a carrier for HPA it has the following advantages: it improves the adsorption of pollutants and improves photocatalytic activity; it increases the specific surface area of the composite, and it improves the physicochemical properties of HPA. Currently, HPA/SiO_2 nanocomposites are in urgent need of solutions to increase the response to visible light and improve photocatalytic activity.

4. HPA/ZrO_2 nanocomposites

Compared to TiO_2, ZrO_2 has a wider band gap (5.0 eV) and much lower photocatalytic performance. However, Devassy Biju et al [23] made HPA/ZrO_2 nanocomposites using ZrO_2 as a carrier for the catalytic synthesis of alkylbenzenes, and the results showed that the catalytic activity was enhanced by loading HPA onto ZrO_2.

Jiang Chunjie et al [24] loaded Ti-substituted Keggin-type HPA ($Li_5PW_{11}TiO_{40}$, $K_7PW_{10}Ti_2O_{40}$) onto ZrO_2 to produce a series of HPA/ZrO_2 nanocomposites. The photocatalytic degradation of naphthol blue-black was found that the photocatalytic activity of the HPA/ZrO_2 nanocomposites was significantly higher than that of the monomeric HPA, and the high specific surface area of the composites and the electron transfer to Keggin-type HPA were the main reasons for the increased photocatalytic activity. Qu Xuesong et al [25] prepared a PW_{12}/ZrO_2 nanocomposite, and photocatalytic degradation experiments showed that the PW_{12}/ZrO_2 nanocomposite could effectively degrade methylene blue under UV irradiation, and the degradation rate of methylene blue reached over 90% in 90 min. Farhadi Saeid et al [26] prepared a series of PW_{12}/ZrO_2 nanocomposites with different loadings (5%-20%) by the impregnation method. The photocatalytic degradation experiments of 4-nitrophenol and methylene blue showed that the photocatalytic activity of 15% PW_{12}/ZrO_2 nanocomposites was significantly higher than that of HPA and ZrO_2 alone. Salavati Hossein et al [27] investigated the degradation of methylene blue, rhodamine B, Congo red, bromothymol blue, alizarin GG, etc. by HPA/ZrO_2 nanocomposites in an aqueous solution The photocatalytic activity of the organic dye pollutants was investigated. The experimental results showed that the HPA/ZrO_2 nanocomposites had higher photocatalytic performance compared with pure HPA or pure ZrO_2. The photocatalytic mechanism was found that: the specific surface area of HPAZrO_2 nanocomposites was higher than that of HPA or ZrO_2; electron transfer from ZrO_2 to HPA delayed the photogenerated hole-electron complex. Therefore, the photocatalytic activity of HPA/ZrO_2 nanocomposites is significantly higher than that of HPA or ZrO_2. HPA and ZrO_2 are compounded together mainly through electrostatic interaction and hydrogen bonding, avoiding secondary pollution of the environment by HPA. HPA loading on ZrO_2 causes a red shift in the absorption spectrum of the composite, which in turn reduces the band gap of the composite and improves the utilization of visible light. Various HPA/ZrO_2 nanocomposites for organic dye degradation are listed in Table 3.

Table 3. Recent outstanding results of the HPA/ZrO_2 nanocomposites for organic dye degradation

Photocatalyst	Synthetic method	Contaminants	Degradation efficiency	Reference
$K_7PW_{10}Ti_2O_{40}/ZrO_2$	Sol-gel method	Naphthol Blue Black	~90% in 60min	[24]
PW_{12}/ZrO_2	impregnation method	Methylene Blue	~90% in 90min	[25]
PW_{12}/ZrO_2	impregnation method	Methylene Blue	~90% in 120min	[26]
PW_{12}/ZrO_2	Sol-gel method	Methylene Blue Congo red Rhodamin B Bromothymol Blue	99% in 50min 95% in 50min 97% in 50min 72% in 50min	[27]

5. HPA/Ta_2O_5 nanocomposites

Ta_2O_5 has a band gap of 3.9 eV and exhibits some photocatalytic activity only under UV irradiation. Jiang Shujuan et al [28] produced HPA/Ta_2O_5 (HPA= $H_3PW_{12}O_{40}$ and $H_6P_2W_{18}O_{62}$) nanocomposites with different loadings of HPA (5.6-15.3%) by hydrolyzing $TaCl_5$, which exhibited high photocatalytic activity under visible light The HPA/Ta_2O_5 (HPA=$H_3PW_{12}O_{40}$ and $H_6P_2W_{18}O_{62}$) nanocomposites were effective in degrading salicylic acid and rhodamine B under visible light irradiation and still had high photocatalytic activity after three reuses. The structural analysis showed that HPA was uniformly dispersed within the Ta_2O_5 skeleton, effectively reducing the band gap and improving the photocatalytic activity. Currently, HPA/Ta_2O_5 nanomaterials have been widely used in organic synthesis and biodegradation, but little research has been reported in the field of photocatalytic degradation of wastewater.

6. HPA/ CNTs nanocomposites

Carbon nanotubes (CNTs) have been widely used in the field of electrochemistry due to their excellent heat transfer performance, special electrical and optical properties. CNTs can adsorb HPA on the surface of CNTs through chemical adsorption, and produce a synergistic effect, thus enhancing the electrocatalytic performance of HPA/CNTs nanocomposites, which has attracted extensive attention [29]. However, HPA is the only photocatalytic material in HPA/CNTs nanocomposites, so the application of HPA/CNTs nanocomposites in the field of photocatalysis has received little attention. Lisw et al. [30] prepared three-component Au/HPA/CNTs nanocomposites with visible light photocatalytic activity by using Keggin type PW_{12} as a reducing and bridging molecule between Au metal nanoparticles and carbon nanotubes. It was found that the degradation rate of Rhodamine B by Au/HPA/CNTs nanocomposites was about 91% after 2 h visible light irradiation, while the degradation rate of Au/HPA was only 45%. The photocatalytic degradation activity of three-component Au/HPA/CNTs nanocomposites was significantly higher than that of a single component or two-component photocatalysts.

7. HPA/ CdS nanocomposites

As one of the semiconductor materials, CdS has a band gap of about 2.42 eV, which can absorb less than 520 nm light and has a wide response range to visible light. However, in practical applications, CdS has the disadvantages of easy recombination of photogenerated carriers and poor photochemical stability, which leads to low photocatalytic quantum efficiency [31]. Xing et al. [32] prepared CdS-PW_{12}-Au ternary nanocomposites by molecular self-loading method. It was found that the photocatalytic degradation of Rhodamine B by CdS-PW_{12}-Au Au could reach 90% under the condition of 0.6% Au nanoparticles loading and 2 h illumination. Besides, HPA can be used not only as reducing, encapsulating, and bridging molecules but also as cocatalyst in photocatalysis.

8. Conclusion and outlook

The synergistic interaction between HPA and nanoparticle carriers can improve the physicochemical properties of HPA/nanocomplexes, such as high specific surface area, more active sites, lower photogenerated electron-hole complexation rate, etc., which in turn significantly enhance the photocatalytic performance of HPA-based nanocomplexes. However, there are still many problems to be solved and further researched in terms of theoretical and practical applications: (1) the preparation method of HPA-based nanocomplexes needs to be further improved, which in turn leads to uniform-sized nanoparticles and improved photocatalytic activity; (2) the construction of multifaceted HPA-based nanocomplexes and the improvement of the stability of HPA-based nanocomplexes to avoid the leaching of HPA, which leads to secondary pollution; (3) the HPA-based nanocomplexes for photocatalytic degradation of printing and dyeing wastewater, most of which choose dyes as the target molecules for degradation. To further determine the photocatalytic performance of HPA-based nanocomplexes, difficult pollutants can be used as degradation or reaction substrates; (4) At present, the photocatalytic research of HPA-based nanocomplexes mostly stays in the laboratory stage, and the technology should be further optimized in the future in terms of preparation methods and recycling, to truly realize the industrialization of nanotechnology.

9. Acknowledgments

This work was supported by the Science and Technology Research Program of Chongqing Municipal Education Commission (grant No. KJQN 201804501).

10. References

[1] Omar M. L. Alharbi, Al Arsh Basheer, Rafat A. Khattab and Imran Ali 2018 J. *Mol. Liq.* **263** 442.
[2] Z. Liu, Y. Liu, G. Zeng, B. Shao, M. Chen, Z. Li, Y. Jiang, Y. Liu, Y. Zhang and H. Zhong 2018 *Chemosphere* **203** 139.
[3] Dandan Zhu and Qixing Zhou 2019 *Environ. Nanotechnol. Monit. Manage.* **12** 100255.
[4] Dasari Ayodhya and Guttena Veerabhadram 2018 *Mater. Today Ener.* **9** 83.
[5] Fang He, Zhenxing Wang, Yuexiang Li, Shaoqin Peng and Bin Liu 2020 *Appl. Catal.,* B **269** 118828.
[6] C. Karthikeyan, Prabhakarn Arunachalam, K. Ramachandran, Abdullah M. Al-Mayouf and S. Karuppuchamy 2020 *J. Alloys Compd.* **828** 154281.
[7] James J. Walsh, Alan M. Bond, Robert J. Forster and Tia E. Keyes 2016 *Coord. Chem. Rev.* **306** 217.
[8] Hai-Lei Cao, Feng-Ying Cai, Hai-Bo Huang, Bahar Karadeniz and Jian Lü 2017 *Inorg. Chem. Commun.* **84** 164.

[9] Solomon Omwoma, Christopher T. Gore, Yuanchun Ji, Changwen Hu and Yu-Fei Song 2015 *Coord. Chem. Rev.* **286** 17.
[10] Giuseppe Marcì, Elisa I. García-López and Leonardo Palmisano 2014 *Eur. J. Inorg. Chem.* **2014** 21.
[11] Dandan Li, Pengtao Ma, Jingyang Niu and Jingping Wang 2019 *Coord. Chem. Rev.* **392** 49.
[12] Lin Yang, Yuan Shen, Yongzhen Chen, Xiang Pan, Xiuli Wang and Xiang Wang 2019 *Inorg. Chem. Commun.* **108** 107493.
[13] Kazuya Nakata and Akira Fujishima 2012 *J. Photochem. Photobiol., C* **13** 169.
[14] H. Jin, Q. Wu and W. Pang 2007 *J. Hazard. Mater.* **141** 123.
[15] S. Tangestaninejad, M. Moghadam, V. Mirkhani, I. Mohammadpoor-Baltork and H. Salavati 2008 *Ultrason. Sonochem.* **15** 815.
[16] Changjun Yang, Lihong Tian, Liqun Ye, Tianyou Peng, Kejian Deng and Ling Zan 2011 *J. Appl. Polym. Sci.* **120** 2048.
[17] H. Shi, T. Zhang, T. An, B. Li and X. Wang 2012 *J. Colloid Interface Sci.* **380** 121.
[18] Chencheng Dong, Jiahui Ji, Zhe Yang, Yifei Xiao, Mingyang Xing and Jinlong Zhang 2019 *Chin. Chem. Lett.* **30** 853.
[19] T. Rasheed, M. Adeel, F. Nabeel, M. Bilal and H. M. N. Iqbal 2019 *Sci. Total Environ.* **688** 299.
[20] Shengyang Tao, Yuchao Wang, Yongxian Yu, Yonglin An and Wenping Shi 2013 *J. Environ. Chem. Eng.* **1** 719.
[21] Yang Hou, Jicheng Ma, Ting Wang and Qiang Fu 2015 *Mater. Sci. Semicond. Process.* **39** 229.
[22] H. Li, S. Gao, M. Cao and R. Cao 2013 *J. Colloid Interface Sci.* **394** 434.
[23] B. Devassy, F. Lefebvre and S. Halligudi 2005 *J. Catal.* **231** 1.
[24] Chunjie Jiang, Yihang Guo, Changwen Hu, Chungang Wang and Danfeng Li 2004 *Mater. Res. Bull.* **39** 251.
[25] Xuesong Qu, Yihang Guo and Changwen Hu 2007 *J. Mol. Catal. A: Chem.* **262** 128.
[26] Saeid Farhadi and Masoumeh Zaidi 2009 *Appl. Catal., A* **354** 119.
[27] H. Salavati, N. Tavakkoli and M. Hosseinpoor 2012 *Ultrason. Sonochem.* **19** 546.
[28] S. Jiang, Y. Guo, C. Wang, X. Qu and L. Li 2007 *J. Colloid Interface Sci.* **308** 208.
[29] Shiwen Li, Xuelian Yu, Guangjin Zhang, Ying Ma, Jiannian Yao and Pedro de Oliveira Carbon 2011 *CARBON* **49** 1906
[30] Shiwen Li, Xuelian Yu, Guangjin Zhang, Ying Ma, Jiannian Yao, Bineta Keita, Nadjo Louisc and Hui Zhaod 2011 *J. Mater. Chem.* **21** 2282.
[31] Azam Khan, Zia-ur-Rehman, Muneeb-ur-Rehman, Rajwali Khan, Zulfiqar, Amir Waseema, Azhar Iqbal and Zawar Hussain Shah 2016 *Inorg. Chem. Comm.***72** 33.
[32] Xiaolin Xing, Rongji Liua, Xuelian Yua, Guangjin Zhang, Hongbin Cao, Jiannian Yao, Baozeng Ren, Zhenxi Jiang and Hui Zhao 2013 *J. Mater. Chem. A* **4** 1488.

IOP Conf. Series: Materials Science and Engineering 1133 (2021) 012002 doi:10.1088/1757-899X/1133/1/012002

^{131}I radioactive self-expandable metallic stents for the therapy of malignant biliary obstruction

Meirong Liu[1], Mengfei Hou, Qinghe Wu, [2]Chunfu Zhang*
Shanghai Jiao Tong University, Shanghai 200030, China.
[1]meirong_l6@sjtu.edu.cn; [2]cfzhang@sjtu.edu.cn

Abstract. Biliary stent implantation has become a standard approach for patients with unresectable malignant biliary obstruction. Unfortunately, in-stent restenosis after stent implantation still occurs because of tumor overgrowth. The majority of studies show that radioactive stents harbor both the obstructed biliary tract-dilating and tumor-suppressing functions. In this study, we designed a kind of novel ^{131}I-labeled radioactive self-expandable metallic stent (SEMS) that combines the mechanical dilation forces induced by stent expansion and ^{131}I intraluminal brachytherapy, which was prepared by first coating the SEMS with poly (diallyl acrylate) (PDA) and then labeling with iodine-131.The designed stents would not only have a longer patency time but also is able to inhibit biliary compression caused by tumor growth. The PDA coating was robust and flexible. The ^{131}I labeling rate was about 60%, and after a 16-day test of radiochemical stability of ^{131}I, the radioactive stability was around 80%. In addition, the radiochemical stability of ^{131}I could be improved by PDA recoating post-^{131}I labeling. In conclusion, we designed and fabricated a kind of novel ^{131}I-labeled radioactive self-expandable metal stents by labeling ^{131}I on the PDA-coated stent. We provide a new idea of radioactive stents design, which would extend the patency time of the stent while reducing the rate of stent restenosis after implantation.

1. Introduction

Malignant biliary obstruction (MBO) is a common disease in clinic, which is commonly caused by cholangiocarcinoma and pancreatic cancer as well as metastatic disease. Cholangiocarcinoma (CCA) is a primary hepatic malignancy, including intrahepatic, perihilar, and distal extrahepatic tumors. Among them, hilar cholangiocarcinoma accounts for approximately 50% of CCA [1]. MBO usually inhibits bile from the liver into the duodenum, which is accompanied by symptoms such as painless jaundice and pruritus. Unfortunately, due to the lack of obvious early symptoms, MBO is commonly diagnosed at advanced ages. As a result, only 10%-20% of MBO patients can accept resection [2], [3]. Even following surgical resection, the 3- and 5-year survival rates remain low (18-52% and 5-31%, respectively) [4], [5]. For these patients, biliary stent implantation has become an important palliative treatment to palliate symptoms such as pruritus, pain, jaundice and cholangitis [6]. However, there are many cases of tumor ingrowth, tumor overgrowth, epithelial ingrowth, and biliary clogging. Moreover, conventional stents can only facilitate drainage but have no antitumor effect. Consequently, the frequency of stent restenosis is about 50% within six months after stent implantation, which has become a critical factor negatively affecting the prognosis of cancer patients [7]. For this reason, it is essential to prepare a kind of radioactive stent that serves both tumor treatment and drainage to increase stent patency duration and the survival rate of patients.

With the development of interventional radiology, intraluminal brachytherapy (ILBT) provides new opportunities for the therapy of MBO. There are three main methods for the design of existing radioactive stents:(1) ^{125}I seed strands; (2) "Bundled" biliary stent; (3) Integrated ^{125}I seed stent. In

2018, Chen [8] et al. developed an integrated ^{125}I seed stent. They demonstrated that inserting a radioactive biliary stent is more effective in reducing jaundice symptoms, inhibiting tumor growth and prolonging the stent patency, as well as prolonging patient survival when compared to a conventional stent in patients with unresectable MBO. However, the above stents also have obvious disadvantages: they are expensive, additionally, only several radioactive seeds attached to stent, resulting in poor homogeneity of radiation and low efficacy.

PDA has received wide attention for biomedical applications due to its biocompatibility and ease of modification. In 2007, inspired by mussels, Lee [9] et al. discovered that diallyl acrylate can be oxidized and spontaneously self-polymerized under alkaline conditions, and then forms PDA on virtually any surfaces, including noble metals, oxides, polymers, semiconductors and ceramics. In addition, thanks to the catechol group on the PDA surface, PDA can be used for many functions according to different purposes, making it an extremely versatile platform.

Intraluminal brachytherapy provides a high localized dose of radiation to the tumor and reduces the volume of irradiated normal tissues. Therefore, it is highly favored in the treatment of biliary obstruction. Iodine-131 is a β-decay radionuclide with a half-life of 8.02 days. It emits β-rays (99%) and γ-rays (1%). The maximum energy of β-rays is 0.6065 MeV and the main γ-ray energy is 0.364 MeV. ^{131}I has broad application in clinical radiation therapy. For example, Zhao [10] et al. coated single-walled carbon nanotubes (SWNT) with PDA and further modified it with polyethylene glycol. Then, the radionuclide ^{131}I was labeled on SWNT@PDA-PEG, combining tumor nuclear imaging and radioisotope therapy.

Herein, a novel radioactive stent was developed involving PDA coating followed by radionuclide I^{131} labeling. PDA coating was ultra-thin (nm level), and there was no PDA degradation within 60 days when incubated at 37 °C in Dulbecco's modified Eagle's medium (DMEM) + 10% fetal bovine serum (FBS), which demonstrated PDA could remain stable over a long period. Moreover, the PDA-coated stent was labeled with iodine-131 for ILBT, which exhibited excellent radiolabeling yields and radiochemical stability. Our study provides a new idea for radioactive stents design, then reduces the rate of stent restenosis after implantation as well as extends the patency time of the stent.

2. Materials and methods

2.1. Synthesis of PDA coated stent

The stents were cleaned by sonication in isopropyl alcohol and deionized water for 10 min each. PDA-coated stent was prepared by dipping the SEMS in diallyl acrylate solution (2 mg/mL of diallyl acrylate in 10 mM of Tris Buffer solution at pH = 8.5). Then, the reaction mixture was exposed to the air and stirred for 4 h at room temperature.

2.2. Synthesis of iodine-131 labeled stent

The stents were soaked overnight in 5 mM $NaIO_4$ solution and then washed with deionized water for 5 minutes. Iodine-131 was labeled by the chloramine T method. Chloramine T oxidizes iodine ions into iodine or high-valent iodine ions. To label PDA-coated stent with radioisotope iodine-131, iodine-131 (100 μCi) and non-radioactive sodium iodide (150 mCi) and chloramine T (0.88 mg) was added into PBS (0.5 mL, pH=7.5), the PDA-coated stent was then put and shook for 10 min. The radiolabelling yield was measured by a radioactivity meter (HD-175A, Heyiqitx, Beijing, China).

2.3. Recoated with PDA and $NaIO_4$ oxidation after labeling iodine-131

The stent was recoated with PDA (2 mg/mL for 1 h) after labeling, and the coating procedure was the same as above. Then used 5 mM $NaIO_4$ solution to oxidize the outer PDA overnight.

2.4. Radiochemical stability of iodine-131 labeled stent

To examine the radiochemical stability of the final ^{131}I radioactive self-expandable metallic stents under physiological conditions, the ^{131}I radioactive SEMS were immersed into DMEM + 10% FBS for different periods of time (1、2、4、8、16 day). The radioactive stability was calculated by dividing the radioactivity retained on the ^{131}I radioactive SEMS to that of sample.

2.5. Statistics
All experiment data expressed are the calculated mean ±standard deviation (SD). A student's t test was used for differences of statistically significant evaluation, and $p < 0.05$ was regarded as significant differences between groups.

3. Results and discussion

3.1. Preparation of PDA-coated stent
PDA coating was synthesized according to the methods described previously[9]. Specifically, PDA-coated stent was prepared by dipping the SEMS in diallyl acrylate solution for different time (2 mg/mL of diallyl acrylate in 10 mM of Tris Buffer solution at pH = 8.5). Then, the reaction mixture was exposed to the air and stirred for 4 h at room temperature. Diallyl acrylate solution then turned to brown, and finally to black while polymerization was occurring with prolonged time (Figure 1). It was seen from Figure 2 that the surface of PDA-coated stent lost its original metallic luster and turned to black. The experimental results initially demonstrated that the PDA was successfully coated on the stent.

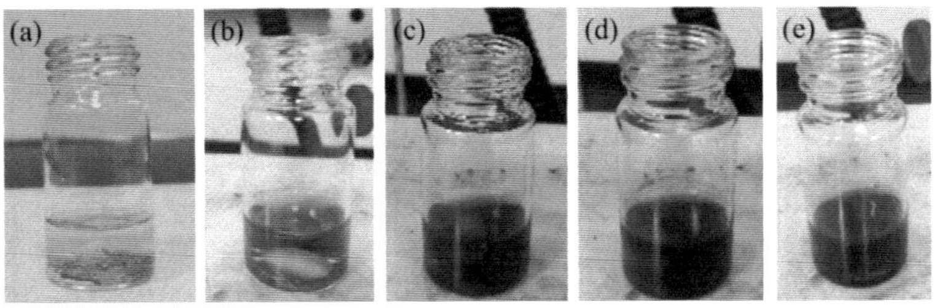

Figure 1. The color change of diallyl acrylate solution after stirring for different time: (a) 0 min; (b) 20 min; (c) 40 min; (d) 1 h; (e) 2 h.

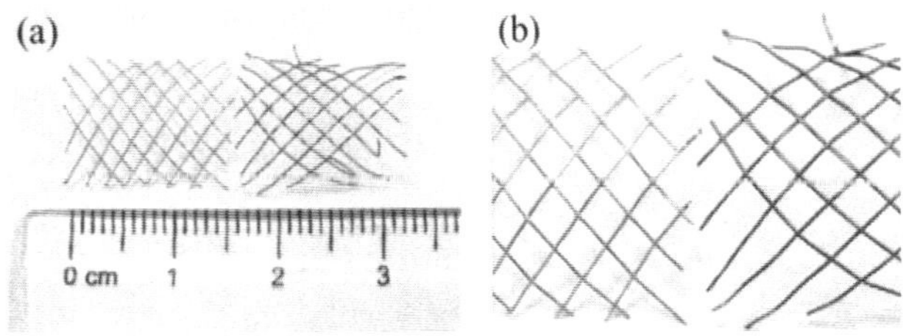

Figure 2. Physical map of the stent. (a) Stent size (left: bare-metal stent; right: PDA-coated stent); (b) The surface color contrast of bare-metal stent (left) and PDA-coated stent (right).

3.2. XPS characterization of PDA coating
As shown in Figure 3, compared to the bare stent, the XPS results of PDA-coated stent showed that the PDA coating generated the characteristic carbon (phenolic) and nitrogen peaks (amine group) (C1s peak at around 285 eV and O1s peak at 532 eV). Meanwhile, XPS spectra in the bare stent group revealed the presence of Cr, Mo elements. However, it should be noted that no metallic Cr or Mo were observed in the XPS spectra after PDA coating, indicating the formation of a polymer coating of 10 nm or more in thickness, which was consistent with the previous report[9]. These peaks indicated successful coating of PDA on the stent. We also characterized the stent after labeling iodine using XPS. From its XPS spectrum, it can be seen that in addition to the peak of PDA, a characteristic binding energy peak of iodine appears, suggesting that iodine could be labeled onto the stent through the PDA coating. The nitrogen-to-carbon signal ratio (N/C) is 0.114 and 0.115 for PDA coated stent and iodine labeled stent, respectively, which is similar to that of the theoretical value (N/C = 0.125), implying that the coating is derived from diallyl acrylate polymerization.

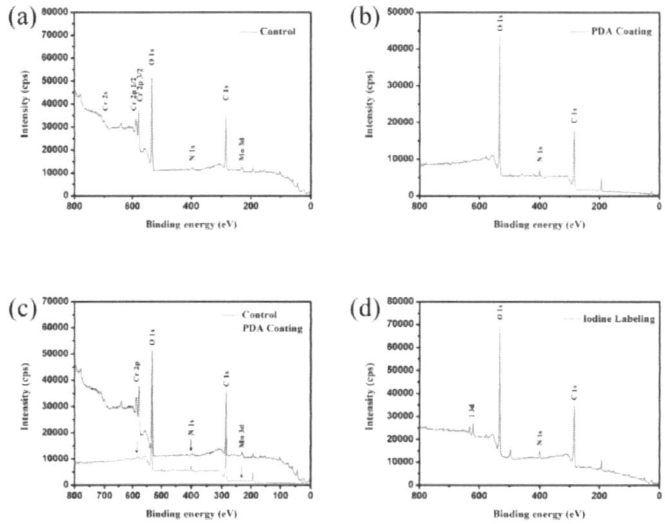

Figure 3. XPS full spectrum. (a) XPS spectrum of bare stent; (b) XPS spectrum of stent coated with PDA; (c) XPS spectrum comparison of bare stent (Control) and stent coated with (PDA g; (d) XPS spectrum of the stent after cold labeling with iodine.

3.3. SEM characterization of PDA coating

To evaluate the flexibility of PDA coatings, we bend the PDA-coated stent and observed the bent part by SEM. The results were displayed in Figure 4 and indicated that even undergoing severe bending, there are no breakages and peeling off in the PDA coating, which implied the PDA coating had excellent flexibility. At the same time, the PDA-coated stent has a rougher surface due to the presence of a coating while the bare stent struts have a smoother surface appearance.

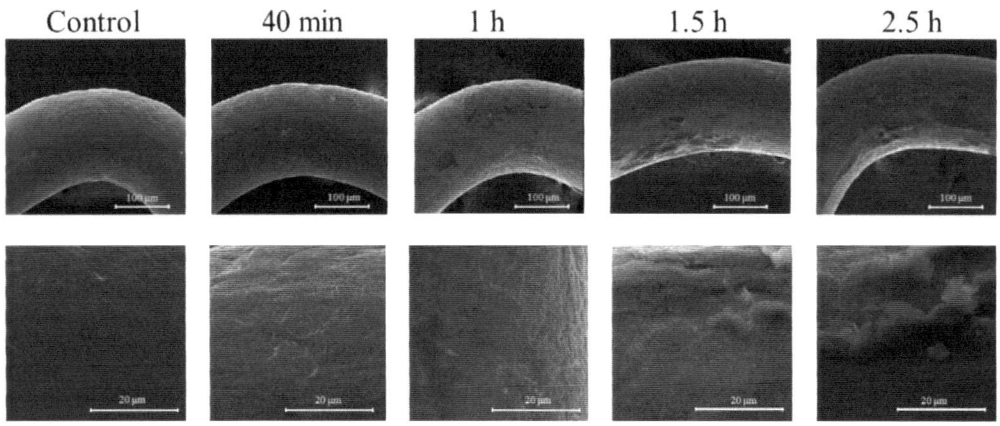

Figure 4. The surface morphology of the stent after being immersed in 2 mg/mL diallyl acrylate for (a) 40 min, (b) 1 h, (c) 1.5 h, (d) 2.5 h.

Since multiple coatings of PDA will be conducted to enhance the radiolabeling stability, the stent was coated with PDA for several times (2 mg / mL for 1 h), then characterized by scanning electron microscopy. As shown in Figure 5, even after four times of PDA coating, the stent did not exhibit any shedding or coating breakages after severe bending. We assessed the surface morphology of PDA-coated stent in comparison to an uncoated bare stent via SEM. The results showed that an increase in surface roughness was found after the coating deposition, which implied the PDA coating had excellent flexibility. Collectively, our results indicated that PDA coating has excellent stability and flexibility, which could adapt to the severe deformation process during biliary stent implantation.

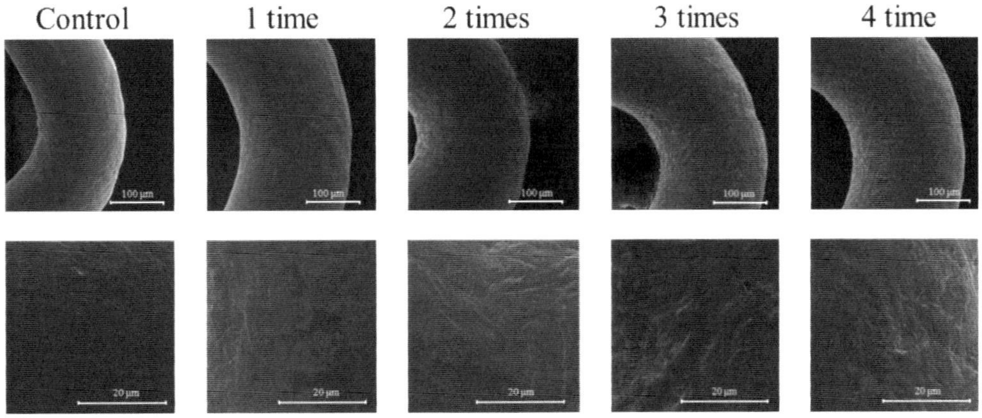

Figure 5. Scanning electron micrographs of the stent immersed in 2 mg/mL diallyl acrylate for 1 hour each time, coated with PDA 1-4 times.

3.4. Degradation of PDA coating

To examine whether the PDA coating will be degraded, PDA-coated nanoparticles were prepared, then PDA coated nanoparticles were incubated in DMEM + 10% FBS. The degradation of PDA was monitored by transmission electron microscope by measuring the thickness of the coating layer (About 200 nanoparticles). As shown in Figure 6, the thickness of PDA coating after immersing in DMEM+10% FBS for 60 days nearly remained unchanged compared with the initial thickness (about 32 nm). The experimental results demonstrated that there were no PDA degradation reactions occurring, indicating that the PDA coating had good stability.

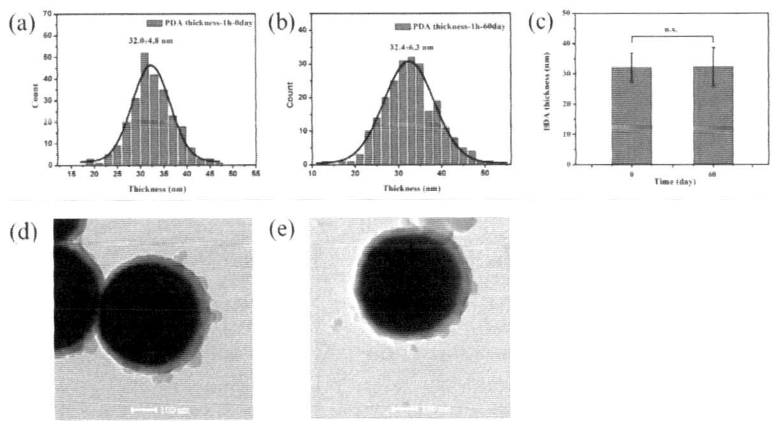

Figure 6. Thickness distribution map of PDA coating. (a) The initial thickness of PDA; (b) the thickness of PDA after immersing in DMEM+10% FBS for 60 days; (c) Significant difference analysis of PDA; (d) Transmission electron micrograph of PDA-coated nanoparticles (initial); (e) Transmission electron micrograph of PDA-coated nanoparticles after immersing in DMEM+10% FBS for 60 days.

3.5. Labeling and characterization of iodine-131

Iodine-131 was labeled by the chloramine T method. The labeling efficiency of iodine-131 and the radiochemical stability in DMEM+10% FBS were evaluated. The iodine labeling efficiency was 50%, and it was increased to 60% when further oxidization of PDA with sodium periodate ($NaIO_4$), making it further crosslinked.

3.6. PDA coating and oxidation after labeling

To improve the radiochemical stability of iodine-131, the stent was recoated with PDA after labeling, and the coating procedure was the same as above. Results showed that further PDA coating achieved moderate improvement for radiochemical stability of I^{131}, which was about 10% higher than the stents that did not wrap PDA again (Figure 7). The radiochemical stability of iodine-131 was then further improved by using sodium periodate to oxidize the outer PDA, which was more stable than the conventionally physical absorption method [11]. These results indicated that the radiochemical stability of iodine-131 was further improved after oxidation of the outer PDA coating with $NaIO_4$.

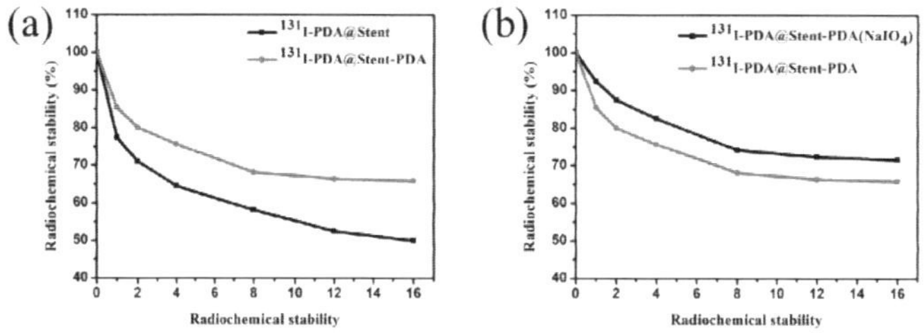

Figure 7. Radiochemical stability of iodine-131. (a) Change of radiochemical stability of ^{131}I after recoated with PDA after the previous labeling; (b) Change of radiochemical stability of ^{131}I after oxidation of the outer PDA coating with $NaIO_4$.

4. Conclusion

In summary, we fabricated a novel ^{131}I-labeled radioactive self-expandable metal stents. PDA was self-polymerized under alkaline conditions on the surface of the stent. Iodine-131 was labeled onto the PDA coating by electrophilic attack on the ortho carbon of the catechol group of PDA. In addition, radiolabeling yields and radiochemical stability of ^{131}I were further improved by recoating with PDA and then oxidation with $NaIO_4$. The radioactive stent had excellent radiochemical stability and could achieve long-term tumor-killing purpose. Therefore, we provided a new method for radioactive stent preparation, which would reduce the rate of stent restenosis after implantation and extend the patency time of the stent.

5. Acknowledgement

This study was supported by the Open Research Fund from National Facility for Translational Medicine (Shanghai) (TMSK-2020-135), Innovation Research Plan supported by Shanghai Municipal Education Commission (2019-01-07-00-02-E00064) and the National Natural Science Foundation of China (81571729).

6. References

[1] Blechacz B 2017 *Gut Liver* Cholangiocarcinoma: Current Knowledge and New Developments **11** 13-26.

[2] Fernandez Y Viesca M and Arvanitakis M. 2019 *Clin Exp Gastroenterol* Early Diagnosis And Management Of Malignant Distal Biliary Obstruction: A Review On Current Recommendations And Guidelines **12** 415-32.

[3] Kanno Y, et al. 2019 *J Gastrointest Cancer* EUS-Guided Biliary Drainage for Unresectable Malignant Biliary Obstruction: 10-Year Experience of 99 Cases at a Single Center. **50** 469-77.

[4] Zhu H D, et al. 2012 *J Hepatol* A novel biliary stent loaded with ^{125}I seeds in patients with malignant biliary obstruction: preliminary results versus a conventional biliary stent **56** 1104-11.

[5] Wang H W, Li X J, Li S J, Lu J R and He D F. 2021 *World J Clin Cases* Biliary stent combined with iodine-125 seed strand implantation in malignant obstructive jaundice **9** 801-11.

[6] Yun J H, Jung G S, Park J G, Kang B C, Shin D H, Yun B C and Lee S U 2016 *Cardiovasc Intervent Radiol* Malignant Hilar Biliary Obstruction: Treatment by Means of Placement of a Newly Designed Y-Shaped Branched Covered Stent **39** 582-90.
[7] Shatzel J, et al. 2016 *World J Gastrointest Endosc* Drug eluting biliary stents to decrease stent failure rates: A review of the literature **8** 77-85.
[8] Chen W, Fang X M, Wang X, Sudarshan S K P, Hu X Y and Chen H W 2018 *J Xray Sci* Technol Preliminary clinical application of integrated ^{125}I seeds stents in the therapy of malignant lower biliary tract obstruction **26** 865-75.
[9] Lee H, Dellatore S M, Miller W M and Messersmith P B 2007 *Science* Mussel-inspired surface chemistry for multifunctional coatings **318** 426-30.
[10] Zhao H, et al. 2016 *Theranostics* Polydopamine Coated Single-Walled Carbon Nanotubes as a Versatile Platform with Radionuclide Labeling for Multimodal Tumor Imaging and Therapy **6** 1833-43.
[11] Kim Y H, et al. 2011 *Small* Tumor targeting and imaging using cyclic RGD-PEGylated gold nanoparticle probes with directly conjugated iodine-125 **7** 2052-60.

IOP Conf. Series: Materials Science and Engineering 1133 (2021) 012003 doi:10.1088/1757-899X/1133/1/012003

Numerical simulation analysis of fluid-solid coupling of water-iubricated bearings

Yong Liu[1a], Xingsheng Lao[1], Chunhui Dai[1] and Shiwei Yao[1.]
[1] Wuhan Second Ship Design and Research Institute, NO.19, Yangqiaohu Avenue, Jiangxia District, Wuhan, Hubei, China
[a]seasmile2004@qq.com

Abstract. The numerical simulation analysis of the fluid-solid coupling of the ship's boring water-lubricated bearing is carried out to study the influence law of the inclination and eccentricity of the rotating shaft on the pressure distribution and fluid flow state in the bearing. The results show that the tilt of the rotating shaft changes the shape of the inner flow passage of the water-lubricated bearing, resulting in unstable flow in the bearing. When rotating shaft is in normal, the pressure value in the water-lubricated bearing decreases first and then increases in the axial direction. However, after the rotation of the shaft is tilted, the change of the pressure value is the opposite; the stress of the rubber layer generally shows an increasing trend with the increase of the inclination angle.

1. Introduction

Water-lubricated bearings have the advantages of low cost, simple structure and convenient maintenance. Using water as working medium can save a lot of strategic resources such as oil and precious non-ferrous metals, and can fundamentally avoid oil pollution to water resources. It has broad application prospects in ships, hydraulics and deep-sea machinery [1] [2]. Therefore, water-lubricated bearings have attracted a great deal of attention as important objects in the fields of environmental science, materials science and mechanical engineering. However, due to the particularity of its structure and the complexity of the operating environment, it is relatively difficult to use finite element software for simulation research. Most of the current researches focus on analyzing its elastohydrodynamic lubrication mechanism and discussing it as a steady state problem. Little attention is paid to the flow field distribution characteristics in the gap of water-lubricated bearings under non-steady-state conditions. Wang Jiaxu et al. [3] proposed a research method for numerical simulation of water-lubricated bearings. Lu Lei et al. [4] carried out two-dimensional numerical simulation of six-groove water-lubricated rubber alloy bearings and obtained some meaningful conclusions. Wang Youqiang et al. [5] carried out numerical calculations on eight longitudinal groove water-softened rubber bearings, and obtained the variation law of friction coefficient with speed and load. Liu Yu et al. [6] and Tian Yuzhong [7] studied the influence of internal water tank structure on lubrication performance by means of MATLAB.

In this paper, the structure of the water-lubricated bearing of the ship is used as the object, and the computational fluid dynamics method is used to study the relationship between the internal pressure distribution characteristics of the water-lubricated bearing, the fluid flow and the groove structure, and the bearing deformation by numerical simulation. Interrelationships, these simulation results have important reference value for guiding material design and structural improvement.

2. Analysis model

The structure of the water lubricated bearing is shown in Figure 1. The rubber bearing is embedded in a steel sleeve. The rubber bearing has an outer diameter of 156 mm and the shaft has a diameter of 100 mm. The bearing material is a new type of polymer material, which is pressed from ultra-high molecular weight polyethylene powder and polyimide. The properties of the polymer materials are shown in the table 1. In the calculation, the shaft speed is 145 rad/s, the friction coefficient between the shaft and the bearing in the seawater is 0.0663, and the shaft inclination angles are 0.0917°, 0.183°, 0.275° and 0.367°, respectively.

Table 1. Properties of the polymer materials.

Density (g/cm^3)	Hardness (HD)	Water absorption rate (%)	Young's modulus(MPa)	Poisson's ratio
0.95	72.8	0.0289	58.3	0.49

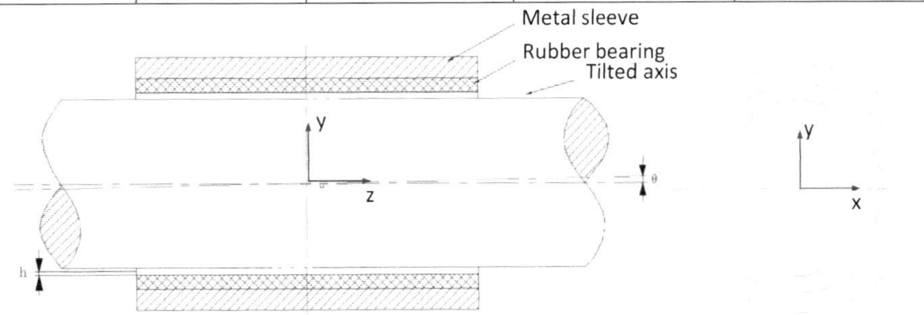

Figure 1. Schematic diagram of water lubricated bearing structure.

3. Results and discussion

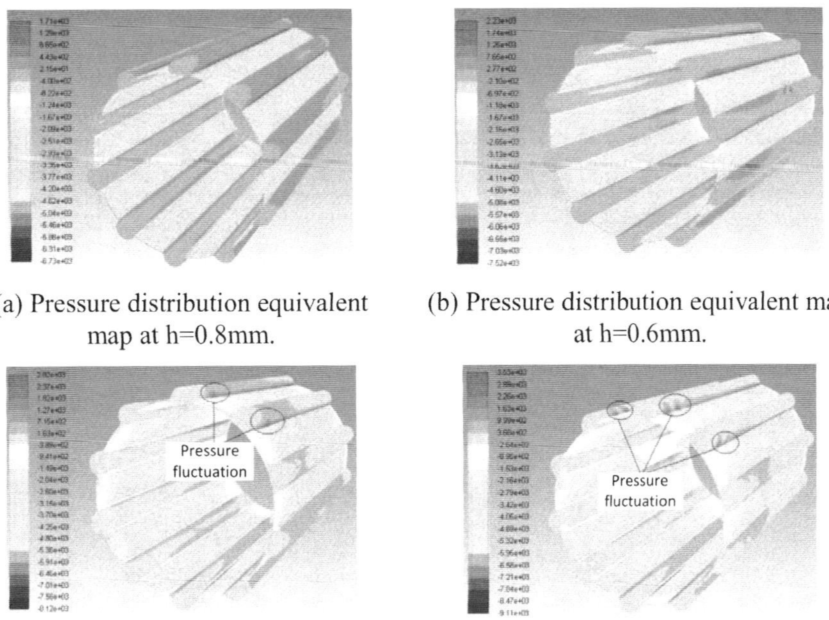

(a) Pressure distribution equivalent map at h=0.8mm.

(b) Pressure distribution equivalent map at h=0.6mm.

(c) Pressure distribution equivalent map at h=0.4mm.

(d) Pressure distribution equivalent map at h=0.2mm.

Figure 2. Equivalent diagram of pressure change at different tilt angles.

The water pressure bearing pressure cloud diagram is shown in Figure 2 at different tilt angles. It can be seen that as the inclination angle increases, the maximum pressure value in the water-lubricated bearing also increases, and the maximum value appears at the groove-to-bearing transition. When h=0.4mm, the upper end of the groove of the water-lubricated bearing fluctuates. When h=0.2mm, there is more obvious fluctuation, which indicates that the flow field fluctuation in the water-lubricated bearing increases with the increase of the inclination angle.

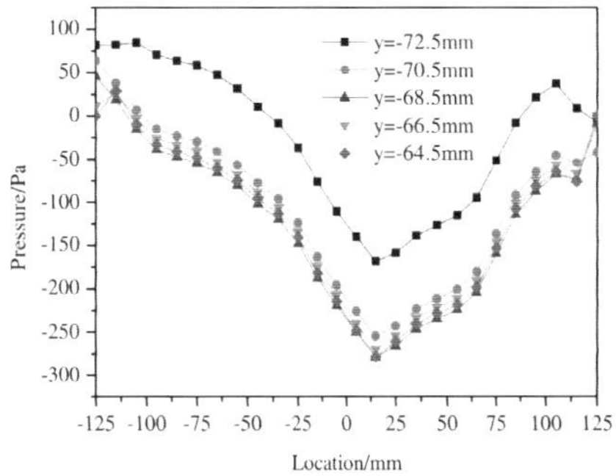

(a) X=0 cross-sectional pressure at h=1.0mm.

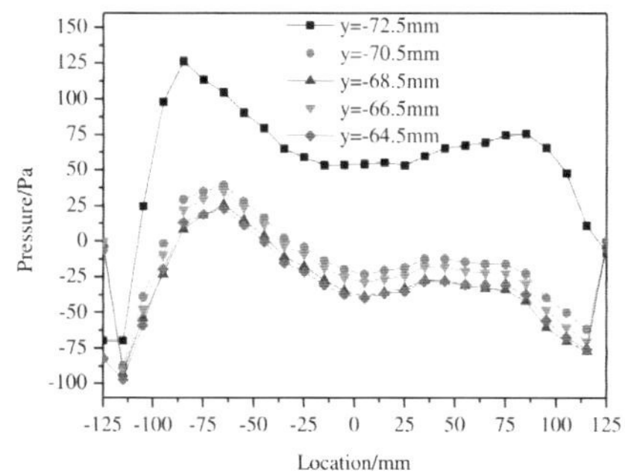

(b) X=0 cross-sectional pressure at h=0.8mm.

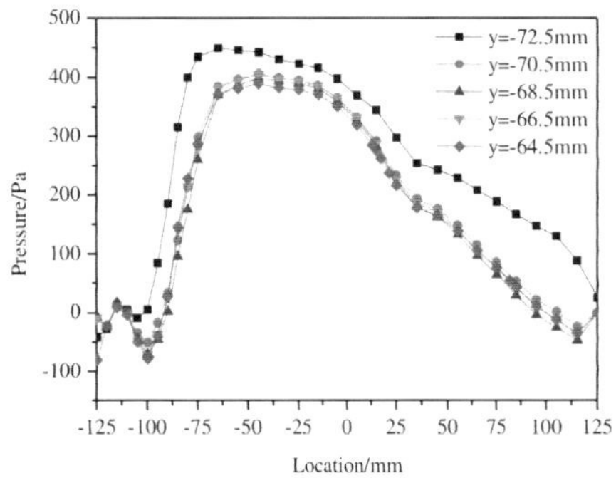

(c) X=0 cross-sectional pressure at h=0.6mm.

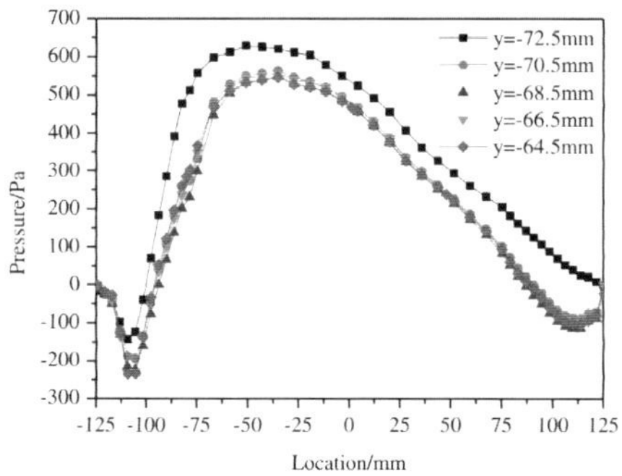

(d) X=0 cross-sectional pressure at h=0.4mm.

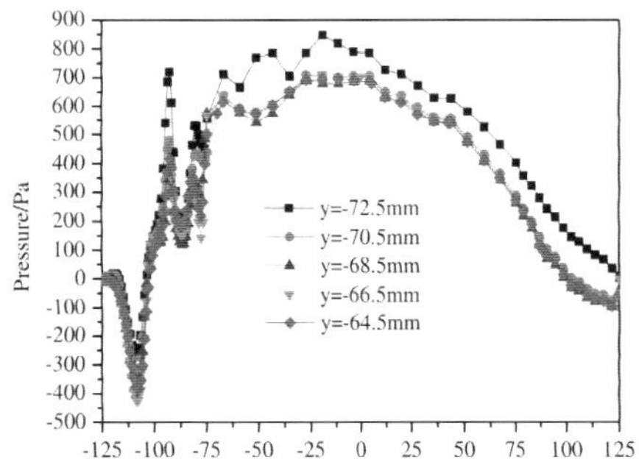

(e) X=0 cross-sectional pressure at h=0.2mm.

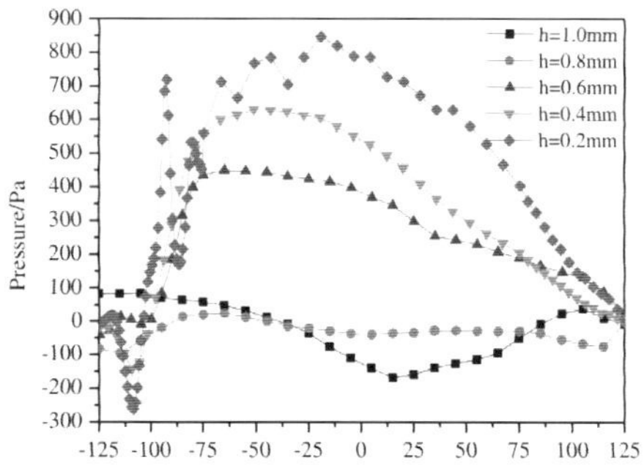

(f) Pressure at the bottom of different h-value trenches.

Figure 3. Pressure changes in the water lubrication x=0 section.

Figure 3(a)-(e) show the pressure changes in the water lubrication x=0 section under different inclination angles of the rotating shaft. It can be seen that in the water-lubricated bearing, the pressure at any radial height is consistent with the change in the axial direction. Moreover, when the rotation axis does not tilt, that is, h=1.0, the pressure value of the bearing at the same radial height decreases first and then increases in the z direction; and when the rotation axis is inclined, that is, h=0.6 mm, h When =0.4mm and h=0.2mm, the pressure values of the bearings at the same radial height increase first and then decrease in the z direction. Figure 3(f) is a summary of the maximum pressure values for different angles of inclination. It can be seen that as the angle of inclination increases, the pressure inside the bearing increases. The maximum pressure value at h=0.2mm indicates that the fluid fluctuation in the bearing is the most severe.

Figure 4 shows the comparison of fluid flow velocity between the shaft and the bearing in the case where the shaft is tilted and eccentric. As can be seen from the figure, the fluid flow velocity grows as the inclination angle of the shaft increases. However, it does not change significantly with the change of the axis eccentricity. At the same level, the fluid velocity in the water-lubricated bearing when the shaft is tilted is greater than the fluid velocity when the shaft is eccentric.

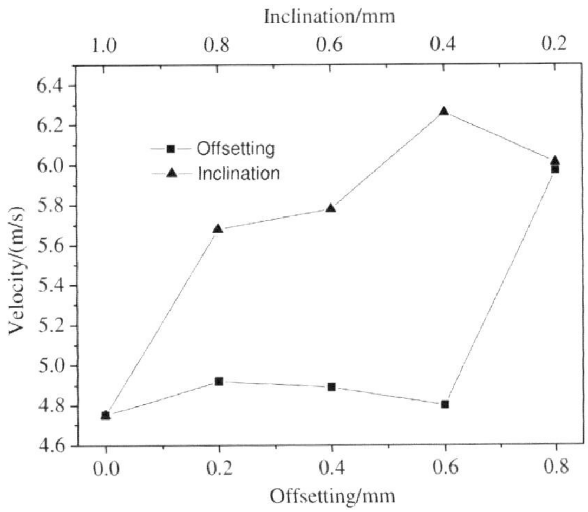

Figure 4. Comparison of maximum speed when the axis is tilted and eccentric.

Figure 5(a)-(d) are the deformation and stress clouds of the water lubricated bearings at different tilt angles of the rotating shaft, respectively. It can be seen that the deformation amount of the rubber layer in the water-lubricated bearing has a maximum value of 1.1475 μm when the inclination is 0.183°, and a minimum value of 0.8695 μm when the inclination is 0.275°. As the tilt angle increases, the amount of deformation of the rubber layer gradually concentrates toward the bottom of the bearing, and the stress of the rubber layer increases first, then decreases and then increases.

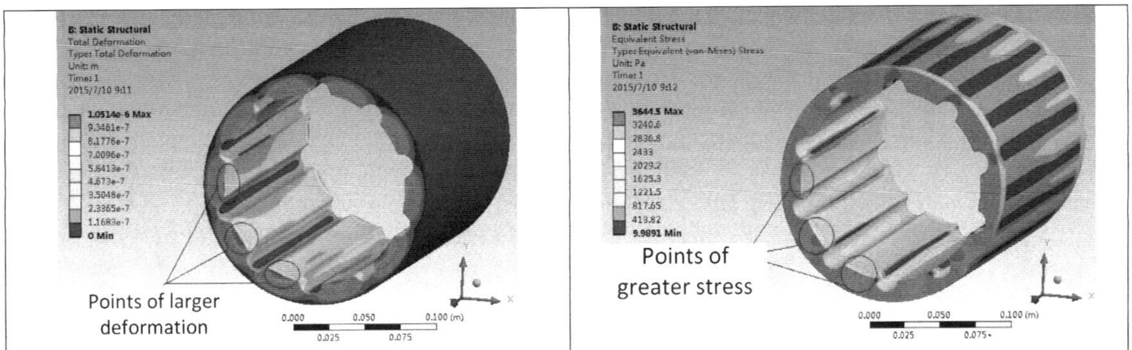

(a) Deformation and stress diagram of water lubricated bearings with shaft tilt of 0.0917°

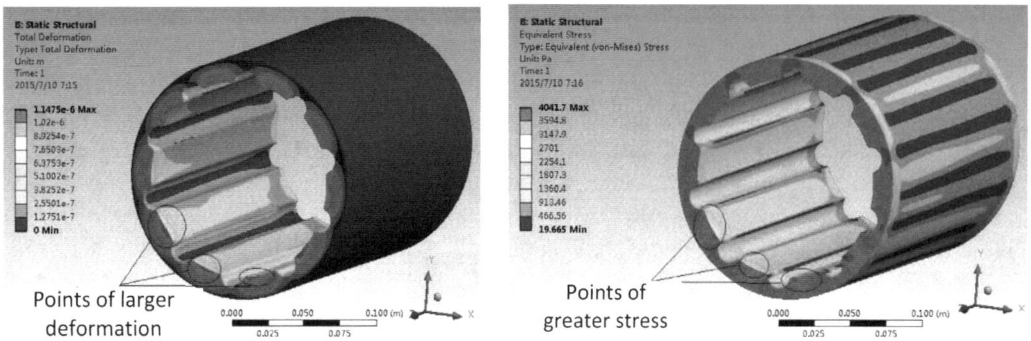

(b) Deformation and stress diagram of water lubricated bearings with shaft tilt of 0.183°

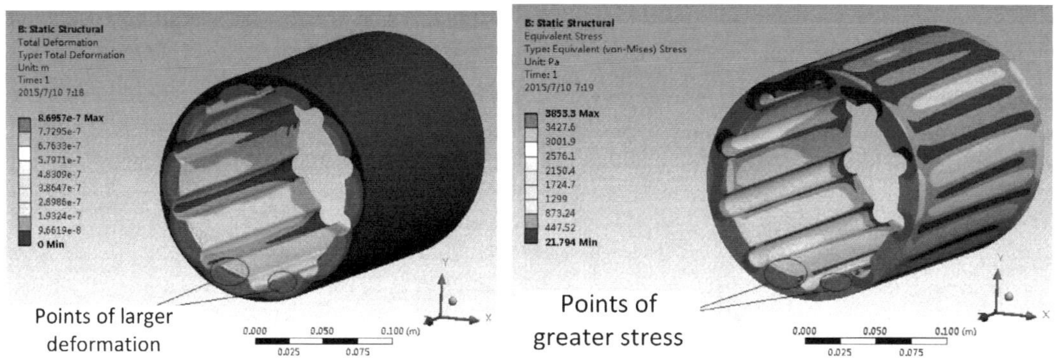

(c) Deformation and stress diagram of water lubricated bearings with shaft tilt of 0.275°

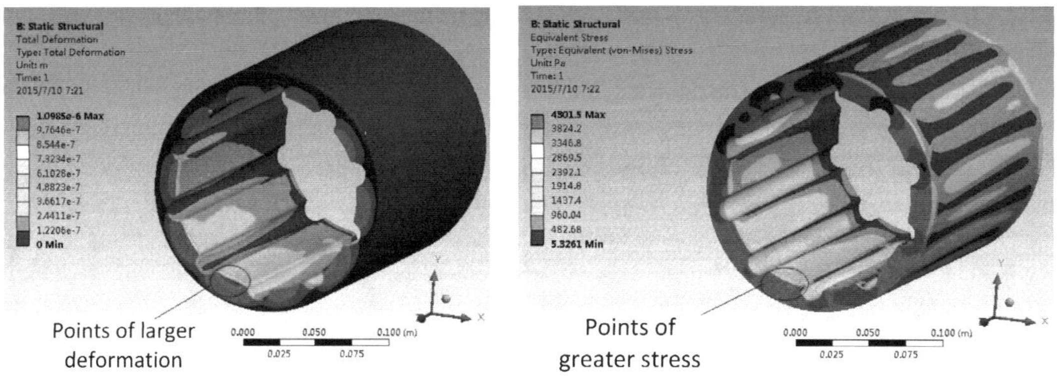

(d) Deformation and stress diagram of water lubricated bearings with shaft tilt of 0.367°

Figure 5. Deformation and stress clouds of the water lubricated bearings

4. Conclusions

Through the numerical analysis of the fluid-solid coupling of the ship's boring water-lubricated bearing, the following conclusions are obtained:

(1) The inclination of the rotating shaft changes the shape of the flow passage of the water-lubricated bearing inner cavity, so that the instability of fluid flow in the bearing increases;

(2) When the shaft is not tilted, the pressure value in the water-lubricated bearing first decreases and then increases in the axial direction, and after the shaft is tilted, the pressure value in the water-lubricated bearing changes in the opposite direction;

(3) Under the same degree, the fluid velocity in the water-lubricated bearing when the shaft is tilted is greater than the fluid velocity when the shaft is eccentric;

(4) The stress of the rubber layer increases first, then decreases and then increases with the increase of the inclination angle. Generally, the stress increases gradually.

5. References

[1] N. S. Tayeb and R. M. Gadelrab. Wear, 192, 112 (1996)
[2] S. Bahadur. Wear, 198, 123 (1994)
[3] J. X. Wang, J. B. Yu and F. Tian. Journal of lubrication and sealing, 4, 62 (2005)
[4] L. Lu, J. X. Wang and K. Xiao. Mechanical Design, 27, 94 (2010)
[5] Y. Q. Wang and C. Z. Yang. Lubrication Engineering, 26, 23 (2001)
[6] Y. Liu, Z. L. Liu and Z. X. Wu. Lubrication Engineering, 34, 71 (2009)
[7] Y. Z. Tian, Z. L. Liu and Y. Liu. Ship & Ocean Engineering, 38, 56 (2009)

IOP Conf. Series: Materials Science and Engineering 1133 (2021) 012004 doi:10.1088/1757-899X/1133/1/012004

A new drilling fluid system for plugging argillaceous shale reservoirs

Fuchang You [1], Man Shu, Yinfu Han
Jingzhou Jiahua Technology Company, Ltd, Jingzhou, Hubei 434000 China
[1]1029415926@qq.com

Abstract. Argillaceous shale reservoir development is mainly based on extended reach horizontal wells, and has complex geological characteristics. In the process of long horizontal drilling, wellbore instability is easy to occur. In this paper, the mechanism of argillaceous shale instability is explored by testing the mineral composition of argillaceous shale. The method of plugging argillaceous shale with nano-sio$_2$ is proposed, and a new nano plugging drilling fluid formulation system is constructed. The system mainly includes specially designed fluid loss additive SFL and EY, flow pattern regulator VBS, lubricant RH, bentonite, nano silica and inorganic salt. Through a series of experimental tests, the rheological property, thermal stability, anti pollution and plugging ability of the system were evaluated. Finally, with the help of SEM, the micro analysis of core profile before and after plugging with nano drilling fluid is carried out, and the plugging mode of nano drilling fluid is further elaborated. The comprehensive results show that nano-sio$_2$ has good compatibility with water-based drilling fluid, and has good effect on inhibiting argillaceous shale expansion and plugging argillaceous shale fractures. It can play the role of stabilizing shale, effectively meet the needs of target block, and provide reference for the follow-up study of plugging argillaceous shale reservoir.

1. Introduction

With the growth of global energy demand and many problems of existing oil and gas resources, energy problem has become a problem faced by many countries. As an unconventional energy, shale gas has been paid more and more attention by various countries. Vigorously developing shale gas plays an important role in alleviating energy pressure. With the increase of shale gas production depth, shale gas reservoir gradually evolved from shale reservoir to argillaceous shale reservoir [1]-[4]. Argillaceous shale is a transitional form between argillaceous rock and shale. Compared with shale, it has deeper burial depth, higher clay content, more microfractures and more abundant shale gas reserves [5]. At present, a large number of researchers have developed various kinds of drilling fluid systems for shale reservoirs, but there are few reports on drilling fluid systems for argillaceous shale reservoirs. Argillaceous shale formation has obvious bedding, developed microfractures and strong water sensitivity. In the process of drilling in long horizontal section, it is easy to fall off and collapse. On the other hand, the horizontal section of argillaceous shale well is relatively long, which further increases the probability of wellbore instability and formation collapse. Therefore, in the process of drilling argillaceous shale horizontal wells, to solve the problem of wellbore instability has become the key to the selection and design of drilling fluid. Oil based drilling fluid system is often used to drill extended reach horizontal wells in argillaceous shale reservoir, but oil-based drilling fluid usually faces the problems of environmental pollution and high cost, while water-based drilling fluid does not contain base oil, so it has natural advantages in environmental protection [6]. Therefore, it is necessary to analyze the causes of argillaceous shale wellbore instability and find a solution for drilling argillaceous shale reservoir with water-based drilling fluid.

Compared with the nanometer size of argillaceous shale, the size of plugging materials in drilling fluid, such as bentonite, barite, film-forming plugging agent and calcium carbonate, is generally more than micron [7]. The pore throat diameter of shale is between 30 ~ 5000nm, and the permeability is low. The traditional drilling fluid additives can not effectively block the micro pores of argillaceous shale, so it can not prevent the free water in the drilling fluid from invading the argillaceous shale [8]-[12] Based on the analysis of physical and chemical properties and wellbore instability mechanism of argillaceous shale, this paper proposes a method of plugging micro fractures in argillaceous shale with nano water-based drilling fluid. A set of drilling fluid suitable for argillaceous shale reservoir is constructed, and a lot of experimental analysis is carried out, which provides a scientific basis and theoretical basis for solving the stability of argillaceous shale reservoir, and has high engineering application value.

2. Experiment

2.1. Materials

Four core samples are taken from Fuling argillaceous shale reservoir, with reservoir depth of 2564m ~ 3046m; chemical substances such as bentonite, NaOH, Na_2CO_3, HCOONa and KCl are purchased from Wuhan Tianli Biochemical Technology Co., Ltd.; drilling fluid additive VBS is a flow pattern regulator, and its main component is a polymer obtained by chemical modification, which is used to increase the viscosity and shear of drilling fluid; SFL is a fluid loss reducer, which is mainly composed of a new type of modified starch. It can form dense mud cake on the surface of wellbore and prevent free water in drilling fluid from invading deep formation; EY is also a fluid loss reducer, but it is a modified cellulose, which can be used together with SFL to minimize the filtration of drilling fluid; RH is a lubricant, and its main component is polyol, which forms particulate emulsion in drilling fluid through cloud point effect They are all provided by Jiahua Technology Co., Ltd. The nano materials used are composed of three different sizes of nano silica, and their mass ratio is 1:1:1. The parameters of three kinds of nano silica are shown in Table 1.

Table 1. Physical parameters of nano silica with different sizes

Material type	Average particle size（nm）	Apparent density（g/cm^3）
NP-SiO2-A	30~60	0.15~0.18
NP-SiO2-B	100~200	0.18~0.2
NP-SiO2-C	300~500	1.8~2.2

2.2. Experiment method and formula

The preparation methods of drilling fluid in this study are in accordance with "GB / T 16783.1-2006 / ISO 10414-12001 and SY / T 7377-2017". The basic drilling fluid formula is as follows:

Water+3%Bentonite+0.2%NaOH+0.2%Na_2CO_3+40%HCOONa+5%KCl+1%VBS+2%SFL+0.5%EY +3%RH+10%Nano plugging agent （1.3g/cm^3）

3. Results and discussion

3.1. Core mineral analysis

Argillaceous shale is a kind of laminated rock with transition form from argillaceous rock to shale. In the process of deposition, it is easy to form imperfect texture and foliation. In the process of geological tectonic movement, argillaceous shale swells with the increase of burial depth, which leads to certain cracks in its structure. Therefore, argillaceous shale generally has the characteristics of large specific surface area, small pores, complex structure, developed microfractures and easy to absorb water and expand. The argillaceous core of shale gas well in Fuling block is analyzed by XRD, and the composition and content of clay minerals and non clay minerals are analyzed, so as to provide valuable information for us to explore the causes of shale instability. The analysis results are shown in Figure 1.

From the geological point of view, it is generally believed that the higher the content of clay minerals in rock, the stronger its hydration expansion; the higher the content of quartz, the stronger its hard brittleness. According to the analysis of argillaceous shale mineral composition, the argillaceous shale minerals in Fuling block are mainly clay and quartz, followed by plagioclase and dolomite, and a small amount of potash feldspar, calcite and pyrite. The content of quartz ranges from 34% to 51%. The high content of hard brittle quartz makes argillaceous shale hard brittle, which adds many factors to argillaceous shale instability. The distribution of clay minerals is between 22% and 47%. The clay minerals are mainly mixed layer and illite, without montmorillonite. The content of hydration dispersion and expansive clay minerals is low. It can be concluded that the hydration dispersion and expansion ability of argillaceous shale are weak, and the main way of wellbore instability is spalling. Therefore, it is necessary to slow down the cracking and shedding of argillaceous shale, so as to improve the wellbore stability.

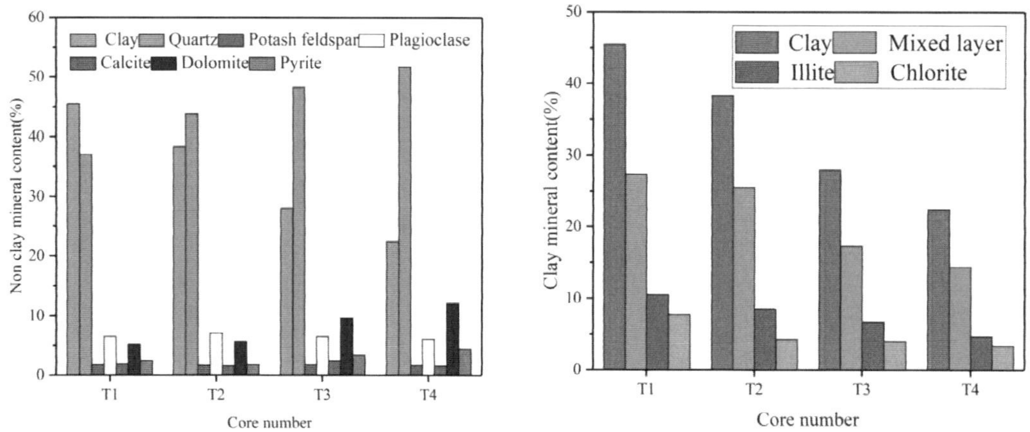

Figure 1. Mineral analysis of argillaceous shale core

3.2. Nano silica plugging argillaceous shale

The pore throat and microfracture size of argillaceous shale are generally between tens of nanometers and hundreds of nanometers. Most of the hard brittle argillaceous shale which is easy to cause wellbore instability contains closed or open bedding and microfracture, and the capillary force is strong. Under the action of downhole positive pressure difference, the free water in drilling fluid is easy to invade the formation, leading to the expansion and cracking of argillaceous shale along the primary fractures and bedding. With the continuous invasion of free water, the fractures continue to develop and expand, eventually leading to wellbore collapse. The particle size of nano materials is generally between 1 ~ 100nm, which is matched with the size of micro pores of argillaceous shale, and can enter into the micro cracks of argillaceous shale for better plugging. Nano silica can be divided into surface modification and no surface modification. The surface modified nano-SiO_2 can disperse well in solution, but it is expensive and not suitable for oilfield chemistry. The unmodified nano-SiO_2 is cheaper, but its dispersibility is poor. On the other hand, a single size of nano materials is difficult to meet the needs of different sizes of cracks, so we use ultrasonic vibration instrument to evenly disperse three different sizes of nano silica in aqueous solution to make nano solution. The inhibition effect of different concentrations of nano solution on argillaceous shale expansion is different. The linear expansion rate experiment is adopted to investigate the inhibition effect of different concentrations of nano solution. The experimental results are shown in Figure 2.

It can be seen from the above figure that before 4h, the concentration of nano solution has little effect on the linear expansion rate of argillaceous shale. After 10h, the argillaceous shale expansion rate of clean water is significantly higher than that of nano solution, which indicates that nano solution can significantly inhibit the hydration expansion of argillaceous shale. The linear expansion rate of argillaceous shale has a positive correlation with the immersion time of argillaceous shale, and a negative correlation with the concentration of nano solution. After 48 hours, the expansion rate of argillaceous shale in each group basically approaches the fixed value. The higher the concentration of nano solution, the stronger the inhibition and the lower the argillaceous shale expansion rate. However, when the concentration of nano particles reaches 60%, the inhibition tends to the maximum, which has little relationship with the concentration. This may be because the volume of microfracture space in argillaceous shale is limited, and excessive nano particles will not enter the microfracture. Therefore, the linear expansion rate of argillaceous shale with 80% and 60% concentration of nano solution has little difference. Considering the preparation cost of nano solution, 60% concentration of nano solution is selected as the plugging agent of shale in the follow-up test of this study.

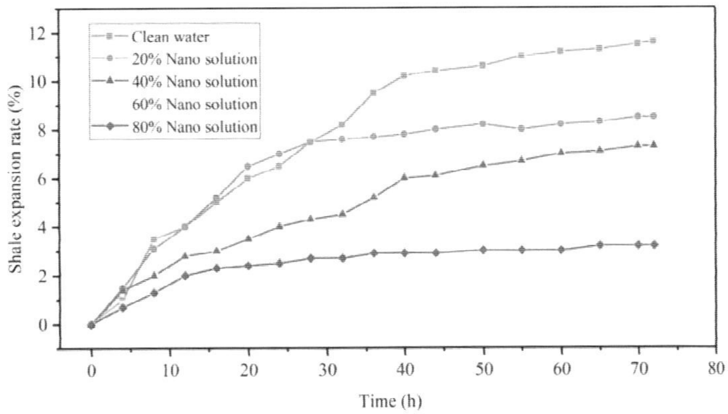

Figure 2. Inhibition of nano solution with different concentrations

3.3. Performance evaluation of nano drilling fluid

4. Rheological properties

The dosage of any kind of drilling fluid treatment agent in drilling fluid is limited to a certain extent. It is not that the more the dosage is, the better the performance is. However, if the dosage is too small, the performance of drilling fluid can not be improved correspondingly. Generally, the dosage of additives should not only meet the requirements of drilling fluid performance, but also ensure the principle of minimum cost. Therefore, in order to determine the optimal dosage of nano solution in drilling fluid, the effects of different dosage of nano solution from 0% to 30% on the basic properties of drilling fluid were investigated. The experimental results are shown in Table 2.

Table 2. Basic properties of nano water based drilling fluid

Dosage(%)	Status	AV(mPa·s)	PV(mPa·s)	YP(Pa)	Φ6/Φ3	FL_{API}(ml)	pH
0	B-R	23	14	9	5/3.5	6.2	9.6
	A-R	22	14	8	7/6		
5	B-R	22	14	8	4.5/3.5	6.0	9.8
	A-R	24	15	9	10.5/8.5		
10	B-R	23.5	15	8.5	6/5	5.6	9.8
	A-R	27	17	10	10/8.5		
15	B-R	25	16	9	7/5.5	5.8	10.1
	A-R	29	18	11	11/9		
20	B-R	28	17	11	8/6	5.6	10.2
	A-R	33	21	12	11/10		
25	B-R	35	13	12	10/9	5.2	10.2
	A-R	40	25	15	14/12.5		
30	B-R	37	22	15	15/13.5	4.8	10.4
	A-R	41	23	18	16/15.5		

*B-R—Before rolling; A-R—After rolling

It can be seen that the viscosity and water loss of nano drilling fluid increase with the increase of nano API content. It shows that the nano solution has a certain thickening and filtration reducing effect, and the nano material itself is neutral, does not react with water-based drilling fluid, does not cause pH change, and has good compatibility with water-based drilling fluid. However, when the dosage of nano solution exceeds 20%, the viscosity and shear force of drilling fluid increase sharply. In the process of tripping, the drilling fluid with too high viscosity and shear force is easy to produce greater exciting pressure and suction pressure, which leads to wellbore instability. Considering the performance of drilling fluid and the preparation cost of nano solution, the recommended dosage of nano solution is 10%.

5. Thermal stability

With the increase of formation depth, the formation temperature of argillaceous shale formation increases gradually. Therefore, in drilling operation, with the increase of well depth, the temperature resistance of nano drilling fluid is also increasing. The depth of argillaceous shale reservoir is generally between 2000 ~ 3600m, and the reservoir temperature is between 80 ℃ and 130 ℃. The dosage of nano material is controlled at 10% in laboratory. The influence of temperature on nano water-based drilling fluid is investigated by setting different aging temperatures. The results are shown in Table 3.

Table 3. Effect of temperature on properties of nano drilling fluid

Temperature(°C)	Status	AV(mPa·s)	PV(mPa·s)	YP(Pa)	Φ6/Φ3
80	B-R	22	14	8	4.5/3.5
	A-R	24	15	9	10.5/8.5
100	B-R	23	15	8	6/5
	A-R	26	15	11	10/9
120	B-R	24	15	9	7/5.5
	A-R	28	16	12	11/9.5
140	B-R	26	16	10	8/5.5
	A-R	25	16	9	9/8

A complete statistics of drilling fluid performance is made when the temperature changes between 80 ℃ and 140℃. Through comparative analysis, we can find that. At the same temperature, the performance of drilling fluid before and after aging has little change. With the increase of temperature, the rheological value fluctuates in a small range, and the maximum variation of single property is less than 5%. It shows that nano materials basically do not affect the temperature resistance of nano water-based drilling fluid, which is also in line with the properties of nano materials.

6. Anti pollution

In the process of argillaceous shale drilling, the drilling fluid contacts with the rock wellbore for a long time. It is inevitable that part of rock debris will dissolve in drilling fluid, and this part of foreign matter belongs to pollutants relative to drilling fluid. Therefore, when drilling argillaceous shale formation, the requirement of anti pollution of drilling fluid is relatively high. By adding 5% of different types of pollutants into nano drilling fluid, the pollution process of drilling fluid was simulated, and the anti pollution ability of nano drilling fluid was investigated. The experimental results are shown in Table 4.

Compared with the blank group, the addition of pollutants will pollute the basic properties of drilling fluid, and will not affect the pH of drilling fluid. The addition of pollutants results in the increase of viscosity and water loss of drilling fluid. The pollution ability of different kinds of pollutants is different. The pollution effect of bentonite is the strongest and that of inorganic salt is the weakest. However, even in the case of the most serious pollution, the drilling fluid shear can still be maintained between 8 ~ 12Pa, which can fully meet the requirements of drilling fluid suspended cuttings. In general, nano drilling fluid has strong anti pollution ability and great application potential.

Table 4. Anti pollution performance of nano drilling fluid

Contaminants	Status	AV(mPa·s)	PV(mPa·s)	YP(Pa)	Φ6/Φ3	FL_{API}(ml)	pH
Nothing	B-R	23	14	9	5/3.5	6.2	9.6
	A-R	22	14	8	7/6		
Debris	B-R	28	17	12	8/6.5	6.8	9.5
	A-R	24	15	10	10/8		
Bentonite	B-R	30	18	12	9/7.5	7.4	9.6
	A-R	27	17	10	10/9		
NaCl	B-R	23	16	7	5/4	6.4	9.6
	A-R	25	16	9	8/7		
$MgCl_2$	B-R	23	15	8	5/3.5	6.0	9.4
	A-R	24	16	8	8/7		
$CaSO_4$	B-R	24	17	7	6/4.5	6.2	9.5
	A-R	25	15	10	10/9		

7. Plugging property

Argillaceous shale is the rock type with the smallest pore size in sedimentary rocks, and nano pores make plugging extremely difficult. The particle size of conventional plugging materials is generally in the range of micron size, which can only plug micron pores, but can not solve the problem of nano pores. Micron scale materials are often accumulated outside the argillaceous shale fractures, but not deep into the fractures, which can not achieve the plugging effect. Nano materials can only be used to plug micro pores in argillaceous shale. By measuring the pressure changes at the front and back ends of argillaceous shale, the plugging effect of nano drilling fluid on argillaceous shale is evaluated. This study has a good guiding role in evaluating the plugging performance of argillaceous shale, and can help solve the problem of wellbore instability in the process of argillaceous shale drilling. The argillaceous shale inlet pressure is set at 3.5Mpa, and the experimental results are shown in Figure 3.

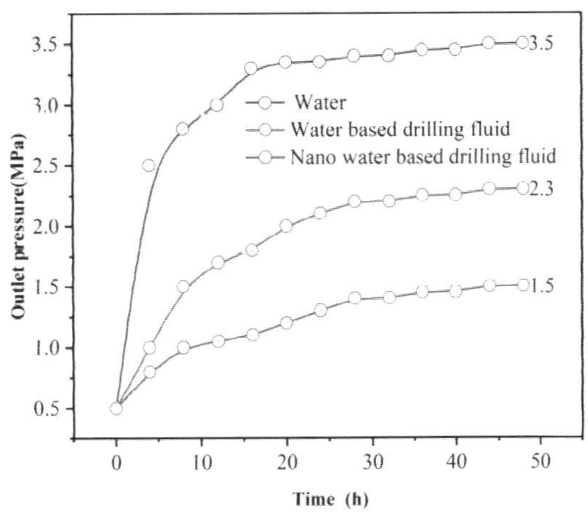

Figure 3. Plugging performance of nano drilling fluid

It can be seen from the curve in the figure above that the plugging ability of clean water to argillaceous shale is poor, and the outlet pressure is close to 3.5Mpa after about 20h. It shows that under the condition of clean water plugging, the pressure of the front section of clean water passes through the argillaceous shale about 20 hours, and the clean water loses its plugging effect. The water-based drilling fluid itself contains a large number of plugging substances. When the water-based drilling fluid is used to plug argillaceous shale, the final outlet pressure of shale is only 2.3MPa; when the nano water-based drilling fluid is used to plug argillaceous shale, the final outlet pressure of argillaceous shale is only 1.5MPa, compared with the water-based drilling fluid, the plugging performance is improved by 34.8%. After the nano materials enter the drilling fluid, the micro fractures of argillaceous shale are effectively blocked, and the transmission of argillaceous shale inlet pressure to outlet pressure is retarded, which indicates that the nano water-based drilling fluid can improve the stability of argillaceous shale.

8. Analysis of core micro morphology

In order to further describe the plugging mode of nano materials, we selected the core profile before and after plugging with nano drilling fluid for SEM analysis, and the plugging time of nano drilling fluid was 48 h. The experimental results are shown in Figure 4.

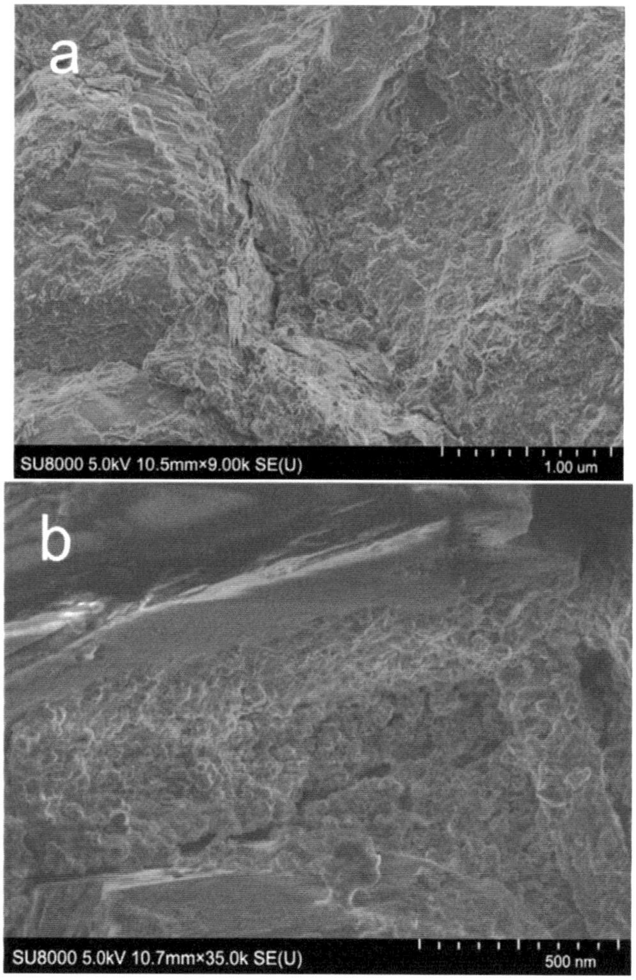

Figure 4. Micro morphology of argillaceous shale

Figure 4 (a) shows the micro morphology of argillaceous shale before plugging with nano drilling fluid. It can be seen from Figure 4 (a) that the mineral bedding of argillaceous shale is obvious, and the micro fractures between different mineral particles are developed, which is not completely dense cementation. It is because of abundant fractures in argillaceous shale that natural gas can be stored in argillaceous shale. However, in the process of shale reservoir exploitation, this kind of microfracture is easy to combine with drilling fluid, resulting in further expansion and amplification of fractures. Finally, it will lead to wellbore instability, and even worse, it will cause the whole well to be scrapped. Therefore, in the process of drilling, it is necessary to add special materials in the drilling fluid to plug these fractures, so as to ensure the safety of the drilling process. Figure 4 (b) shows the micro morphology of argillaceous shale after plugging by nano drilling fluid. It can be seen from Figure 4 (b) that other solid materials in nano drilling fluid are plugged on the fracture surface, but it is difficult to plug nano fractures because their particle diameter is generally larger than 1 μm. However, nano-SiO_2 with different sizes basically gathered at the crack edge. Some nanoparticles even enter into the fracture to repair part of the fracture, shorten the width and length of the fracture, and improve the wellbore stability of argillaceous shale.

9. Conclusions

1. The argillaceous shale reservoir is rich in microfractures, and the clay mineral content is between 22% and 47%. There is no montmorillonite, the hydration expansion ability is weak, and the main way of wellbore instability is spalling. Through a large number of evaluation experiments, the optimal concentration of nano solution is 60%, and the optimal dosage of nano solution in drilling fluid is 10%.

2. The nano drilling fluid system has good rheological properties between 80 ℃ and 130 ℃. And it can resist the pollution of debris, bentonite and inorganic salt. When the inlet pressure is 3.5MPa, the outlet pressure of argillaceous shale sealed with ordinary drilling fluid is 2.3MPa, and the outlet pressure of argillaceous shale sealed with nano drilling fluid is only 1.5MPa. Nano drilling fluid can effectively seal micro fractures in argillaceous shale.

3. The solid particles in common drilling fluid are micron level, most of which are attached to the fracture surface, so it is difficult to plug nano level fractures. Nano silica in nano drilling fluid can enter nano fractures, and improve the stability of argillaceous shale by shortening the width and length of fractures.

10. Acknowledgments

This work was supported by the National Natural Science Foundation of China (Grant No. 519040334, 5180404-4). And the National Major Science and Technology Projects of China (Grant number 2016ZX05060 and 2017ZX05032004-004).

11. References

[1] S. Saleem, M. Vynnycky, H. Metall, Mater. Trans. B 48 (2017) 1625-1635.
[2] Oleg D. Neikov, Mechanical alloying, Handbook of Non-Ferrous Metal Power (Second Edition), (2019).
[3] M. Ramezani, T. Neitzert, J. Achieve. Mater. Manuf. Eng.55/2(2018), p.790
[4] M. Suzuki,Y. Yamaoka, Mater. Trans. 44 (2003) 836-844.
[5] J. Panek, M. Karolus, J. Mater, Sci. Technol (2019).
[6] H. Mizukami, A. Yamanaka, ISIJ Int. 50 (2020) 435-444.
[7] Y. Li, J. Wang, J. Zhang, C. Cheng, Z. Zeng, Proc.36 (2017) 347-357.
[8] E. Wielgosz, T. Kargul, J. Therm. Anal. Calorim. 119 (2015) 1547-1553.
[9] M. Krasnowski, S. Gierlotka, T. Kulik, J. Alloys Compd. 791(2019), p. 75-80.
[10] S. Ueda, K. Kadoi, S. Tokita, H. Inoue, ISIJ Int.7 (2019) 1323-1329.
[11] K. Nova, P. Novak, F. Prusa, J. Kopecek and J. Cech, Metals 9, (2019),p.1.
[12] A. Wedernia, R. Lachheb, J. J. Sunol, J. Saurina, L. Escoda, M. Khitoun, Mater Character 148(2019), p.272.

A new type of low density cement slurry suitable for sandstone formation

Jia Zeng[1,a], Fuchang You[2,b]
[1]Yangtze University College of Technology & Engineering, Jingzhou, Hubei 434020 China
[2]Jingzhou Jiahua Technology Company, Ltd, Jingzhou, Hubei 434000 China
[a]690515926@qq.com [b]1029415926@qq.com

Abstract. Unconsolidated sandstone reservoir is an important unconventional oil and gas resource with the characteristics of porosity, low permeability, strong heterogeneity and many interlayer. The main challenge of unconsolidated sandstone formation is sand production of oil wells, and the more serious situation is high-pressure oil and gas escaping, blowout accidents, environmental pollution and the safety of operators. In order to solve the sensitivity problem of loose sandstone, a low density admixture was developed: 10.5% artificial floating bead + 4.5% power plant floating bead + 35% micro silicon + 30% fly ash + 20% slag. The particle size distribution of single lightweight material is narrow, and the effect of reducing density is not significant. The developed low-density mixing material has wide and uniform particle size distribution, and can reduce the density of cement slurry to the maximum within the minimum dosage. The new low density admixture cement slurry with adjustable density in the range of 1.4 ~ 1.7 g / cm^3 has good rheological property, short thickening time, water loss less than 50 ml, 24 h compressive strength more than 12 MPa, which can meet the needs of site construction. The temperature and pressure test results show that the system has a wide range of temperature and pressure applications and stable performance. Compared with the similar low density cement slurry, it is about 40% lower than the market price and suitable for commercial application.

1. Introduction

Shallow unconsolidated sandstone is an important unconventional oil and gas resource in China, which is characterized by low porosity, strong heterogeneity and many interlayers. It is mainly distributed in Xinglongtai block of Liaohe Oilfield, Xuqian block of Xuzhuang oilfield, Qikou block of Dagang Oilfield, high shallow north area of Jidong Oilfield, northern Songliao Basin and Western Sichuan basin [1]-[3]. The main challenge of shallow unconsolidated sandstone formation is sand production from oil wells. According to the combination mode of formation sand, formation sand is divided into filling sand and skeleton sand [4]. The filling sand is located in the pores of the reservoir. When the velocity of formation pore fluid passes through the sand filling area, the velocity and pressure difference will push the filling sand back out of the well. This stage of sand production is inevitable, and this stage of sand production helps to reduce the flow resistance of pore fluid, improve the formation permeability and increase the oil and gas production. However, when the velocity through the pore exceeds the critical velocity, the pore structure of the weak and loose sandstone is easy to collapse and collapse, causing permanent damage to the reservoir [5]-[7]. The more serious situation is the high pressure oil and gas escaping, blowout accident, pollution of the environment and harm to the safety of operators. Therefore, it is necessary to use cementing slurry to effectively seal the shallow unconsolidated sandstone formation to maintain the long-term stable production of oil wells. In the field of petroleum engineering, low density cement slurry is often used to seal loose sandstone formation [8]-[11].It is considered that low-density cement slurry has the advantages of short setting time and good stability, which can greatly improve the construction safety and cementing quality of shallow wells prone to lost circulation.

2. Experimental equipment and methods

2.1. Experimental equipment

During the preparation of cement slurry, OWC-9360 constant speed agitator was used to mix cement slurry, TG-1220C atmospheric pressure thickener was used to maintain cement slurry at normal temperature, DFC-0712B double cylinder pressurized thickener was used to test the thickening time of cement slurry, HH420 electric constant temperature water bath was used to heat the cement slurry, 2NN-D6 six speed rotary viscometer was used to test the rheological reading of cement slurry, and Y-200 universal universal mechanical testing machine was used to test the slurry thickening time The compressive strength of cement paste was tested, and the HTHP water loss of cement slurry was tested by TG-71 high temperature and high pressure water loss instrument.

2.2. Experimental methods

The preparation and experiment of cement slurry in this study are conducted according to SY /T5504.1-2005《Evaluation Method of Oil Well Cement Additives》、SY/T6544-2003《Performance Requirements of Oil Well Water Slurry》 and API specification Spec-10A. Because the cement slurry contains floating bead lightening agent, which is easy to be damaged under the mixing rate of 12000 r/min, which affects the performance of cement slurry; while the mixing rate of equipment for mixing cement slurry on site is only equivalent to that of indoor 2000 r/min, the indoor mixing rate should be controlled by compromise method, combining API standard and field operation conditions. The indoor configuration method is as follows: at the speed of 4000 r/min, pour the mixed ash into the slurry cup within 15s, and then continue to stir for 45s.

3. Design of cement slurry system

3.1. Design principle of cement slurry

There are many problems in the application of grade g Portland cement in unconsolidated sandstone formation. The most fatal problem is that low density cement stone needs longer curing time under low temperature. In order to overcome the above problems, the most important thing is to improve the hydration rate of cement at low temperature, so that the cement slurry can obtain enough strength in a short time to meet the requirements of subsequent drilling. In order to obtain a better low temperature cement slurry system, we need to start from the following three aspects: 1) using efficient early strength agent to make the cement slurry effectively set in a short time [12]. In order to make Portland cement have strength at low temperature, it is necessary to ensure that the cement slurry can effectively set in a relatively short period of time at low temperature, so as to shorten the hydration induction stage of cement as far as possible [13]. For the deep water low temperature and low density cement slurry system, the high efficient low temperature early strength agent is used to improve the hydration rate of cement slurry and accelerate the development of cement strength; 2) light materials are used to reduce the water cement ratio of cement slurry and improve the initial strength of cement paste; 3) Active filler and inert filler are used to fill the voids between cement particles through physical and chemical action to reduce the permeability of cement slurry [14]-[17].

3.2. Design of cement slurry system

After comprehensive consideration of the density, compressive strength, settlement stability, rheology, thickening time and cost of cement slurry, the low-density mixed material composed of floating beads, micro silica, fly ash and slag is considered to achieve high performance under the condition of low cost.

Through a lot of basic research, the main components of low density admixture (DA) are determined as follows: 10.5% artificial floating beads + 4.5% power plant floating beads + 35% micro silica +

30% fly ash + 20% slag.

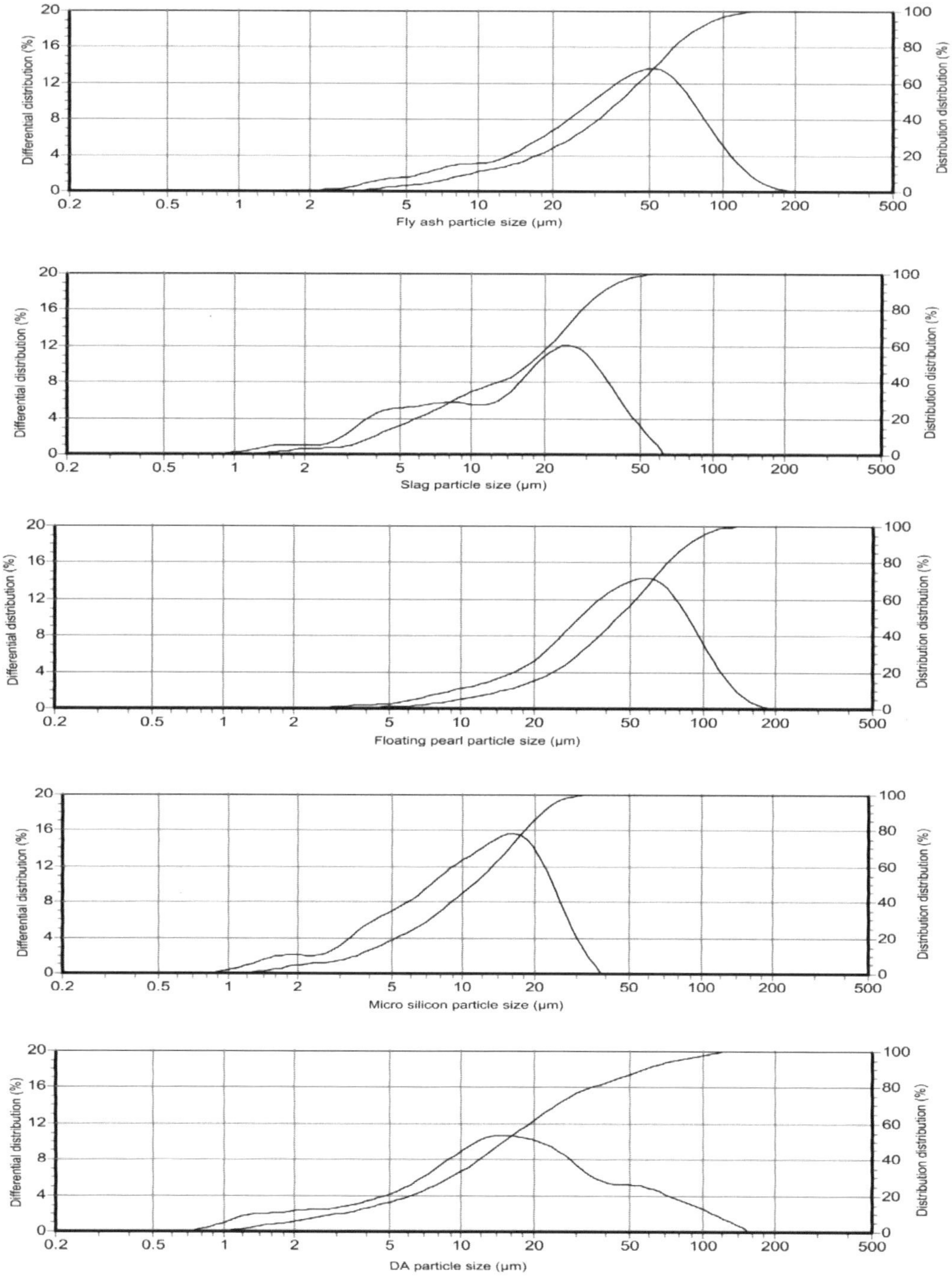

Figure. 1 Particle size of different types of lightweight materials

The main additive materials of cement slurry are early strength agent ACL、dispersant CF42L, fluid loss additive CG81L、retarder RE-L、enhancer ESC-S and defoamer CX60L are used in the preparation of cement slurry. The cement slurry systems with different densities constructed with the above additives are as follows:

0#: 100%Cement+3%ACL+139%Water+1.5%RE-L+4%CF42L+6%CG81L+2%ESC-S+1%CX60L

1#: 0#+91%DA (1.4g/cm3)
2#: 0#+72%DA (1.5g/cm3)
3#: 0#+58%DA (1.6g/cm3)
4#: 0#+32%DA (1.7g/cm3)

4. Results and discussion

4.1. Particle size analysis of low density mixed timber

The compressive strength, density, permeability and water loss performance of low-density cement slurry are mainly based on the particle grading technology. Through the three-level grading principle of solid particles, the primary pores of cement paste are filled and the space network structure of cement paste is improved. Therefore, the particle size distribution of low density mixed materials is very important. The particle size distribution of different types of lightweight materials is shown in Figure 1.

Through the particle size test of single lightweight material and mixed lightweight material, it can be seen that the particle size distribution of different lightweight materials is very different. The results show that the particle size distribution of the mixed low-density materials is enlarged, and the solid particles are evenly distributed in the range of 1-150 μm. This particle size distribution is helpful to improve the spatial structure of cement slurry colloid and enhance the mechanical properties of cement paste.

4.2. Performance evaluation of different density cement slurry system

For the cement slurry system used for cementing, its conventional properties usually include rheological property, stability, thickening time, water loss and compressive strength. These properties are generally concerned by laboratory tests and field cement slurry tests, which are directly related to the construction safety and cementing quality of cement slurry. According to the cement slurry system with different densities, the laboratory evaluated various properties of cement slurry at 50 ℃. The experimental results are shown in Table 1.

Table 1. Rheology and stability of cement slurry with different densities

Number	Φ300	Free water (mL)	Loss water (mL)	Thickening time (min)	Compressive strength (MPa)
1#	173	0	52	309	13.3
2#	156	0	41	273	14.2
3#	149	0	48	280	15.8
4#	142	0	56	284	15.2

It can be seen from the experimental results that: 1) for different density cement slurry formula, its rheological property can meet the cementing requirements, and the cement slurry formula with different density does not have free fluid, and the cement slurry has good stability. 2) The water loss of cement slurry with different density is about 50ml. In order to further control the water loss of cement slurry, more fluid loss additive should be added. 3) The thickening time of cement slurry with different density is between 3 and 7 hours, which can meet the requirements of cementing. 4) Through the evaluation of the compressive strength of different density cement slurry system, it is found that the compressive strength of different density cement slurry system at 50 ℃ is greater than 12MPa, and the compressive strength of cement slurry system can meet the requirements of cementing in offshore unconsolidated sandstone formation.

4.3. Effect of temperature on the performance of cement slurry

The performance of cement slurry in shallow unconsolidated sandstone formation is investigated in laboratory, and the performance variation of cement slurry in the temperature range of 30 ~ 50 ℃ is considered. In order to understand the variation law of cement slurry performance with temperature, the rheological property, thickening property, water loss performance and compressive strength of cement slurry with 1.5g/cm3 density at 30 ℃, 40 ℃ and 50 ℃ were tested in the laboratory. The test results are shown in Figure 2:

The test results show that temperature can significantly affect the basic properties of low density cement slurry, but has little effect on the water loss of low density cement slurry. The performance of cement slurry with density of 1.5g/cm3 at different temperatures can meet the needs of the industry. With the increase of temperature, the cement slurry becomes thinner obviously. The increase of temperature is helpful to promote the flow of cement slurry, so as to improve the dynamic stability of cement slurry [18]-[21]. Low temperature has a negative effect on the hydration reaction of cement slurry, prolonging the thickening time of cement slurry and reducing the initial compressive strength of cement paste.

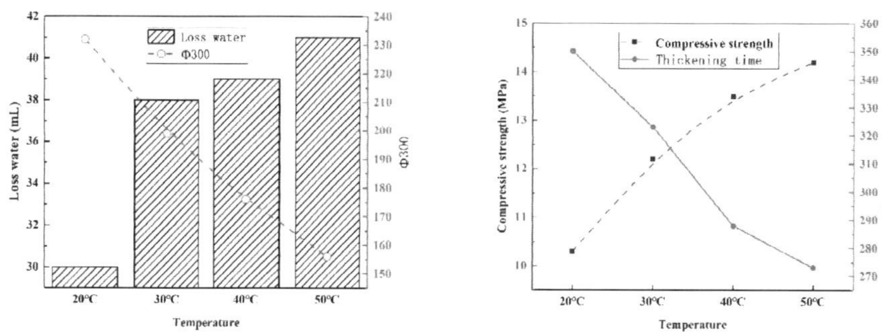

Figure. 2 Effect of temperature on the performance of cement slurry

4.4. Influence of pressure on performance of cement slurry

In order to study the influence of pressure change on the performance of cement slurry, the performance of cement slurry under different pressures of 50 ℃ was evaluated in laboratory. The experimental results are shown in Table 2.

Table 2. Influence of pressure on performance of cement slurry

Test pressure (MPa)	Thickening time (min)	Compressive strength (MPa)
0	312	14.2
10	301	13.8
20	283	14.7
30	268	15.5

The experimental results show that the thickening time of cement slurry and the compressive strength after curing change regularly with the change of test pressure. The compressive strength of cement slurry increases with the time of thickening. It can be seen that the pressure has a great influence on the properties of cement slurry. The reasons for this phenomenon are as follows: (1) some lightening agents are broken, causing changes in the performance of cement slurry; (2) pressure changes the distribution of solids in the slurry. Because the influence of pressure on cement slurry is regular, it is recommended to accurately measure the bottom hole formation pressure in field application. Only by accurately mastering the real pressure data of cement slurry sealing section can we master the variation law of cement slurry according to the experiment and provide better and more direct guidance for field operation.

4.5. Cost comparison of cement slurry

At present, low density cement slurry mostly adopts single lightening material to reduce the density of cement slurry, neglecting the single particle size distribution of lightening material, which is difficult to reduce the density of cement slurry to the maximum extent. Moreover, some lightening materials are easy to cause the cement slurry to separate and disintegrate after curing, such as floating beads. On the other hand, the oil and gas industry is increasingly depressed, and major enterprises are pursuing "reducing cost and increasing efficiency" and seeking better development prospects. The same is true for cementing slurry. Previous researchers have neglected that a single lightening agent is easy to cause high cost of low-density cement slurry. Therefore, this paper has taken this into account at the beginning of the system construction, and on the basis of ensuring the performance of cement slurry, we seek the lowest cost low-density cement slurry. The average price of self-made low-density cement slurry is 2103 yuan / m^3, while the average price of floating bead low-density cement slurry of the same series is 4285 yuan / m3, which is about 40% lower than the market price.

5. Conclusion

The average particle size distribution range of single low density lightening material is narrow, so it is difficult to fill the micropore and reduce the density of cement paste as much as possible. The particle size distribution of low-density composite material is extremely wide and uniform, which can reduce the density of cement slurry in the minimum dosage. The density of the self-developed low-density mixed cement slurry is adjustable, and the density has little effect on the basic performance of the cement slurry, which is more conducive to the field construction. Low density mixed cement slurry is made of inorganic inert materials. Under low temperature and low pressure, the performance of cement slurry is hardly affected by temperature and pressure. Different from deep-water low-temperature low-density cement slurry, the thickening time of low-density cement slurry mixed with loose sandstone can be shortened to 300 min, and the 24 hour compressive strength can exceed 12 MPa, and the average price is about 40% lower than that of deep-water low-density cement slurry.

6. Acknowledgments

This work was supported by the National Natural Science Foundation of China (Grant No.519040334, 5180404-4).And the National Major Science and Technology Projects of China (Grant number 2016ZX05060 and 2017ZX05032004-004).

7. Reference

[1] Hua S D, Yao X, Mechanism of brittleness reduction of oil well cement paste, *Journal of China University of Petroleum*. 26 (2007)108-113.
[2] Shen J Y, Shi L, Li Yong, Analysis and Prospect of cement sheath sealing integrity under large pressure difference, *Natural Gas Industry*. 37 (2017)98-108.
[3] Zhang Y H, Xu M B, Research progress in modification and toughening of cement paste, Applied Chemical Engineering.48(2019) 2198-2202.
[4] Agapiou K, Gamwell R, Influence of recycled rubber tire morphology on the mechanical properties of well cements, ARMA.12 (2016)16-32.
[5] Li Z Y, Guo X Y, Effect of rubber powder on mechanical properties of oil well cement, Petroleum Drilling Technology. 18 (2008) 52-55.
[6] Pan S, Wang Q, Zhan D F, Study on properties and microstructure of cement mortar modified by emulsified asphalt powder, Silicate Bulletin. 38 (2019) 2622-2630.
[7] Liu Y L, Li Z Y, Xue Y T, Study on mechanical properties of water-based epoxy resin cement for cementing, Silicate Bulletin.38 (2019)339-343.
[8] A. Wedernia, R. Lachheb, J. J. Sunol, J. Saurina, L. Escoda, M. Khitoun, Mater Character 148(2019), p.272.
[9] J. Panek, M. Karolus, J. Mater, Sci. Technol (2019).
[10] E. Wielgosz, T. Kargul, J. Therm. Anal. Calorim. 119 (2015) 1547-1553.
[11] M. Ramezani, T. Neitzert, J. Achieve. Mater. Manuf. Eng. 55/2 (2018), p.790
[12] T Qureshi, A Kanellopoulos, A Tabbaa, Construction and Buillding Materials.15 (2018) 192-197.
[13] T Ahn, T Kishi, *Journal of Advanced Concrete Technology*.12 (2018)171-186.
[14] C Joseph, A Jefferson, M Cantoni, *Journal of Advanced Concrete Technology*. 15 (2017)1-8.

[15] Z Yang, J Hollar, X He, Cement and Concrete Composites. 33 (2016) 506-512.
[16] J Zhang, Y Liu, T Feng, Construction and Buillding Materials. 148 (2017) 610-617.
[17] H Xu, J Lian, M Gao, Materials.12 (2019) 13-23.
[18] J Xu, X Wang, Construction and Buillding Materials. 167 (2018) 1-14.
[19] A Muhammad, S Taoutaou, S Shafqat, Materials.3 (2018) 86-92.
[20] J Song, M Xu, W Liu, Advances in Materials Science and Engineering.11 (2019) 35-42.
[21] R Morlat, G Orange, Y Bomal, *Journal of Materials Science*.13 (2017) 4858-4869.

IOP Conf. Series: Materials Science and Engineering 1133 (2021) 012006 doi:10.1088/1757-899X/1133/1/012006

Study on performance improvement of oil well cement slurry by poly ether ether ketone

Fuchang You[1], Jin Tian, Man Shu
Jingzhou Jiahua Technology Company, Ltd, Jingzhou, Hubei 434000 China
[1]1029415926@qq.com

Abstract. After the oil well cement solidifies, the internal original void causes the inherent strength of the cement stone to be insufficient, and the structural integrity of the cement stone is damaged after the external load, which is not conducive to the subsequent oil and gas exploitation. In order to make up for the inherent deficiency of cement paste strength, the effect of PEEK with high strength and high elastic modulus on the properties of oil well cement was studied. The conventional properties and mechanical properties of PEEK cement paste with different contents were evaluated. The microstructure of PEEK cement paste was observed by scanning electron microscope. The experimental results show that compared with the cement paste without peek, the rheological properties of cement paste with 1% ~ 6% peek are worse, the thickening time is shorter, and the water loss is reduced. After curing for 28 days, the compressive strength of 5% peek cement paste decreased by 15%, the flexural strength and impact strength increased by 21% and 11% respectively, and the elastic modulus decreased by 50%. Peek is uniformly dispersed and filled in the cement paste. When the cement paste is subjected to external load, peek will deform, absorb part of the energy and hinder the crack propagation, which improves the ability of the cement paste to resist external load.

1. Introduction

Cementing is a process in which cement slurry is injected into the well wall and casing string from the wellhead. The purpose of cementing is to seal oil, gas and water in different pressure formations and provide guarantee for subsequent construction [1]. After the completion of cementing operation, in the process of oil and gas exploitation, fracturing, perforation and pressure test are often accompanied [2]. These operations often cause uneven stress distribution of cement sheath, produce annulus micro gap, damage the integrity of cement sheath, and cause interlayer sealing failure of cement sheath [3]. Cement sheath is formed after the oil well cement is squeezed into the annulus and solidified. The quality of oil well cement determines the quality of cement sheath, which requires the cement stone researched in laboratory to have excellent resistance to external load.

One of the common methods of strengthening and toughening is to add elastic particles into the slurry. The elastic particles with appropriate particle size can fill the original pores of cement paste and make up for the inherent defects of cement paste. When the cement stone is compressed and squeezed, the particles, as a part of the cement stone skeleton structure, transmit the impact force, deform themselves and absorb energy, so as to improve the mechanical properties of the cement stone. Agapiou K and other studies show that rubber particles can reduce the elastic modulus and improve the impact strength of cement paste. L Zaoyuan [4] studied the effect of hydrophilic rubber powder treated with surfactant on the properties of oil well cement, and determined that the optimal particle size range of rubber powder is 0.15mm ~ 0.18mm, and the optimal dosage is 2%. At the same time, L Zaoyuan [5] also studied the effect of elastic granular material JF on the toughness of cement paste. The 24h compressive strength of 5% JF granular cement paste is 77% higher than that of blank cement paste. P

Shuo [6] studied the effect of emulsified asphalt powder on the mechanical properties of cement paste. When the dosage is 10%, the 28d tensile strength of emulsified asphalt cement paste increases by 20%, the compressive strength decreases by 31%, the flexural strength increases by 49%, and the elastic modulus decreases by 19%. The results of K Nova [7] show that epoxy resin can improve the compressive strength, reduce the elastic modulus and improve the impact strength of cement paste. L Yuliang [8] studied the strengthening and toughening effect of high proportion epoxy resin on cement paste. The results show that the compressive strength of 60% epoxy resin cement paste is 47% higher than that of blank cement paste, the tensile strength is 78% higher, and the elastic modulus is 76.5% lower. These studies show that the proper elastic particles can improve the mechanical properties of cement paste.

Poly Ether Ether Ketone (PEEK) is a kind of special polymer elastic material with high strength, high elastic modulus, chemical resistance and wear resistance, which has been widely used in automobile manufacturing, rail transit, medicine, electronics, aerospace and other industries [9]-[14]. However, there is no report that peek has been used to toughen oil well cement paste. In order to explore the application effect of peek in oil well cement slurry, this paper evaluates the influence of peek on the comprehensive performance of oil well cement slurry, and analyzes the mechanism of PEEK as an external toughening filler to strengthen the mechanical properties of cement paste, so as to provide some reference and guidance for improving the elastic toughness of oil well cement paste.

2. Experiment

2.1. Materials

Water loss agent (Flo-82L) is polyethyleneimine. Its main function is to reduce the amount of water loss of cement slurry. Expansive agent (Bond) is a kind of non-metallic oxide with micro expansion characteristics, which can effectively make up for the volume shrinkage of cement paste. The dispersant (Fsg-S) is sulfonated acetone formaldehyde condensate, the main function is to improve the rheological properties of cement slurry. Reinforcing agent (STR) is a kind of inorganic mineral material with high chemical activity, which can promote the hydration reaction of cement slurry and improve the initial strength of cement paste. The main function of defoaming agent (DFM21) is emulsified silicone oil, which is to eliminate foam in cement slurry. Retarder (H21L) is a mixture of tartaric acid and borate, the main role is to prolong the curing time of cement paste; they all come. From Jingzhou Jiahua Technology Co., Ltd.

Table 1. Basic properties of PEEK

Average grain size /μm	Density /(g /cm3)	Melting point /°C	Tensile strength/MPa	Bending strength/MPa	Impact strength/J/m
50	1.320	334	> 93	140	60~80

Figure 1. The basic microscopic morphology of PEEK

Grade g cement of the Three Gorges project comes from Gezhouba special cement plant. Peek comes from Jilin Engineering Plastics Co., Ltd. The performance parameters of PEEK are shown in Table 1, and the micro morphology is shown in Figure 1.

2.2. Experimental formula design

In this paper, the configuration of cement slurry, the performance test of cement slurry and the performance test of cement paste are carried out in accordance with the corresponding provisions of GB 10238-2005 oil well cement and GB / T 19139-2012 test methods for oil well cement. The curing and testing temperature is 90 °C. The experimental formula is as follows:

1#:100% Grade g cement + 45% Water + 2.5% Flo-82L+ 1.5% STR + 1% Bond + 1% Fsg-S + 0.5% DFM21 + 0.5% H21L

2#: 1#+PEEK

3. Results and discussion

3.1. Basic properties of PEEK cement slurry

In cementing operation, the rheology of cement slurry affects the displacement efficiency of cement slurry, the thickening time determines the height position of cement slurry, and the water loss affects the cementing quality of cement sheath, casing and formation. The conventional properties of cement pastes with different peek dosages were evaluated in laboratory. The experimental results are shown in Table 2 and Figure 2.

Table 2. Rheological properties of PEEK powder cement slurry

Amount/%	φ300	φ200	φ100	φ6	φ3	Δρ/ g/cm3
0	195	130	68	4	3	0.02
1	204	138	65	4	3	0
2	215	144	76	4	3	0
3	220	152	80	4	3	0
4	287	175	122	9	5	0
5	289	233	136	10	7	0
6	300+	223	156	18	16	0

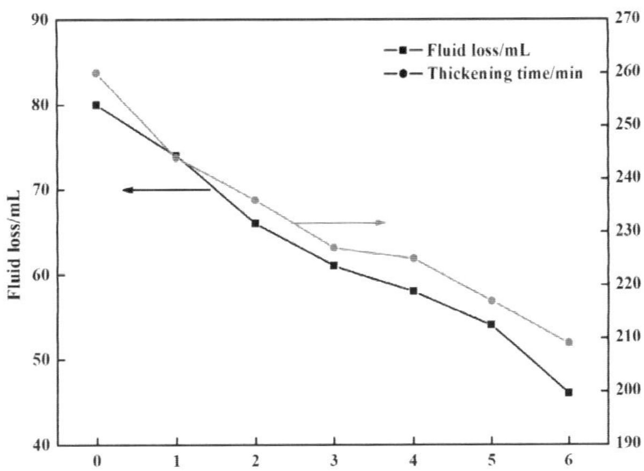

Figure 2. Water loss and thickening time of PEEK cement slurry

The experimental results show that: the settlement stability of cement slurry added with PEEK is good, peek can improve the settlement stability of cement slurry, but peek has an effect on the fluidity of cement slurry. When the dosage of PEEK is less than 3%, it has little effect on the fluidity of cement slurry. With the increase of PEEK, the fluidity of cement slurry becomes worse. When the dosage of PEEK reaches 6%, the reading of φ 300 has exceeded 300. Therefore, peek should be applied in cementing slurry. If the dosage of PEEK is increased, the rheological property of cement slurry should be monitored in time to prevent the high reading of rheological property from making it difficult for cement slurry to pump into the well. Peek can shorten the thickening time of cement slurry, and the thickening time of 6% peek cement slurry is shorter than that of cement slurry without peek 51 min, which is because the void of cement stone containing water is filled with PEEK, the reduction of water greatly shortens the thickening time of cement stone. As can be seen from Figure 2, compared with the blank group, peek can effectively reduce the water loss of cement slurry. When the dosage of PEEK is 6%, the water loss is 46 mL, which is 34 mL less than the blank group. The reason may be that peek is dispersed in cement slurry. When the filter cake is formed, peek fills the narrow pores in the filter cake, making the filter cake more compact and reducing the subsequent water loss.

3.2. Mechanical properties of PEEK cement paste

3.2.1. Compressive strength

The compressive strength of cement paste is the force needed to destroy cement paste per unit area, which is an important parameter in the mechanical properties of cement paste. The compressive strength of cement stone with different dosage of PEEK is tested by universal mechanical testing machine in laboratory. The experimental results are shown in Figure 3.

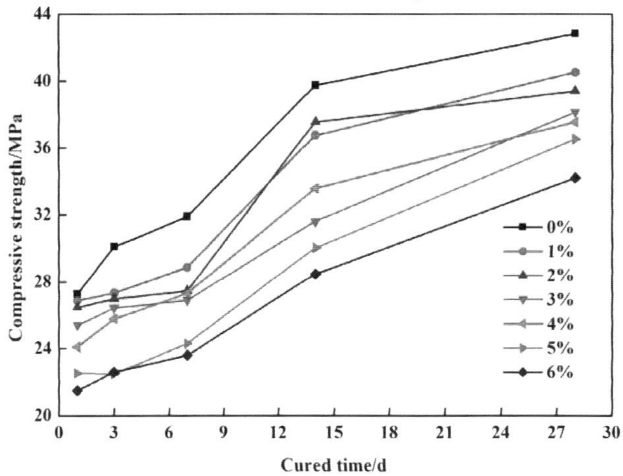

Figure 3. Compressive strength of cement stone at different cured time

After curing for 1 day, the compressive strength of 6% peek cement paste is 21% lower than that of blank cement paste; after curing for 28 days, the compressive strength of 6% peek cement paste is 34.2MPa. Compared with the cement paste without peek particles, the compressive strength of cement paste with 1%, 2%, 3%, 4%, 5% and 6% peek particles decreased by 5%, 8%, 11%, 12%, 15% and 20% after curing for 28 days. Although this does not conform to the principle of material mechanics, it reflects that the filling of elastic materials in cement paste is a double-edged sword. Because peek particles are a kind of foreign material compared with the whole cement paste, the free peek outside the gap will destroy the integrity of the original skeleton structure of cement paste, which is reflected in the mechanical properties, that is, the compressive strength decreases. When the dosage is low, peek particles do not fill the voids in the cement paste and do not share the external load of the cement paste. The loading of PEEK is equivalent to the impurities in the cement paste, which is harmful to the compressive strength of the cement paste. At the optimum dosage, the primary pores of cement paste are basically filled, and the external load is transferred to the interface of the pores. Peek particles are equivalent to a part of the skeleton structure of cement paste, and bear the external load together, with little loss of compressive strength. If the dosage exceeds the optimum, the extra part of PEEK will be harmful to the compressive strength of cement paste as a foreign material, resulting in the decrease of compressive strength of cement paste.

3.2.2. Flexural strength

Flexural strength reflects the ability of cement stone to resist external bending strain, which is a characterization of the toughness of cement stone. The greater the impact strength is, the stronger the ability of object to resist external bending is. The flexural strength of PEEK cement paste was evaluated in laboratory. The results are shown in Figure 4.

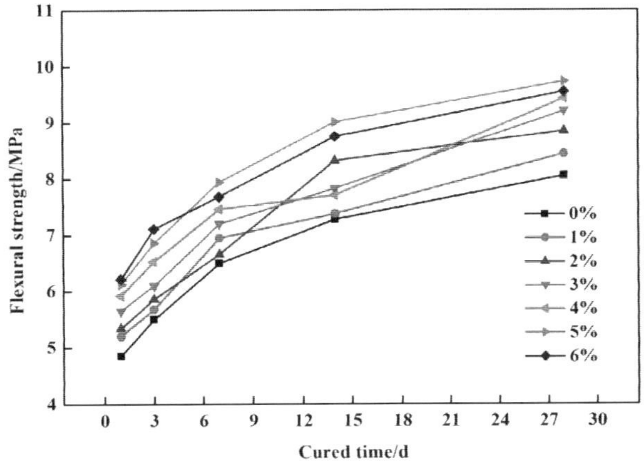

Figure 4. Flexural strength of cement stone at different cured time

It can be seen from the experimental results that peek can significantly improve the flexural strength of cement paste. The longer the curing time, the greater the flexural strength of cement stone. However, it is not that the more peek is added, the better. The flexural strength of 5% peek cement paste develops fastest, and the flexural strength reaches 9.73 MPa after 28 days of curing.

3.2.3. Impact strength

Impact toughness refers to the energy consumed by cement specimen after impact fracture, which can quantitatively characterize the toughening effect of cement paste. The impact strength of PEEK cement paste at different curing time is tested by simply supported beam impact tester indoor. The experimental results are shown in Figure 5.

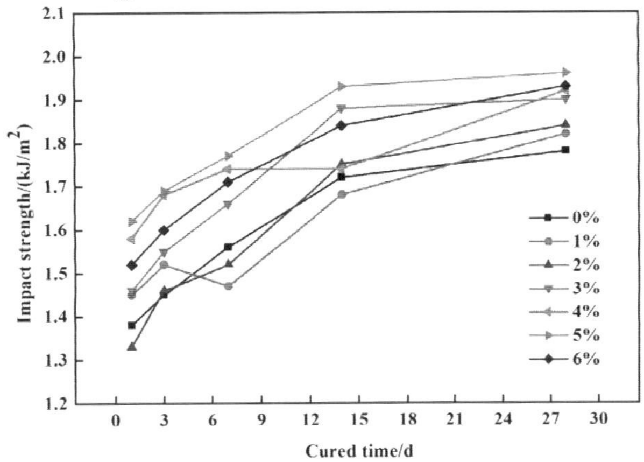

Figure 5. Impact strength of cement stone at different cured time

It can be seen from Figure 5 that peek particles can improve the impact strength of cement paste. With the increase of curing time, the impact strength of cement increases. After curing for one day, the impact strength of 6% peek cement increased by 10% compared with the blank group. After curing for 28 days, the impact strength of six groups of PEEK particles increased by 2%, 3%, 7%, 8%, 10% and 8% respectively. The addition of PEEK particles can improve the impact toughness of cement paste, and the impact strength of 5% peek particles is the highest.

3.3 Stress strain curve of PEEK cement paste

The stress-strain curves of cement paste without peek and cement paste with 5% peek after curing for 1 day were studied in laboratory. The experimental results are shown in Figure 6.

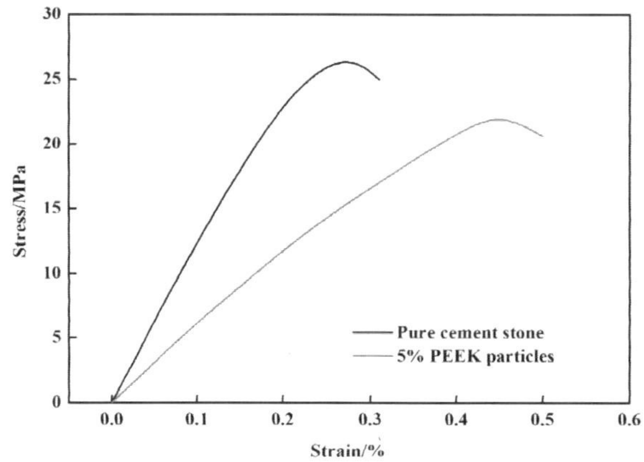

Figure 6. Stress-strain curves of cement stone

According to the stress-strain curve, the elastic modulus of the blank cement paste is 10.3 GPa.. Compared with the blank cement paste, the elastic modulus of the cement paste with 5% peek decreases by 50%, reaching 5.1GPa. It can be seen from Fig. 6 that before reaching the maximum stress, i.e. the cement stone is not damaged, under the same load, the shape of the cement stone added with PEEK changes greatly and the elastic ratio is large. When the maximum stress is reached, i.e. the cement stone is destroyed, the stress reduction rate of the cement stone containing 5% peek is much lower than that of the blank cement stone, and the toughness is relatively large. The reason is that after the cement stone containing 5% peek is destroyed, peek particles have a buffering effect against the pressure exerted by the press by virtue of their excellent elastic toughness, so the stress-strain curve declines slowly compared with the blank group. In conclusion, peek can significantly reduce the elastic modulus of cement paste and improve the elastic toughness of cement paste.

3.4. Toughening mechanism of PEEK cement paste

Figure 7. Micro morphology of blank cement paste

IOP Conf. Series: Materials Science and Engineering 1133 (2021) 012006 doi:10.1088/1757-899X/1133/1/012006

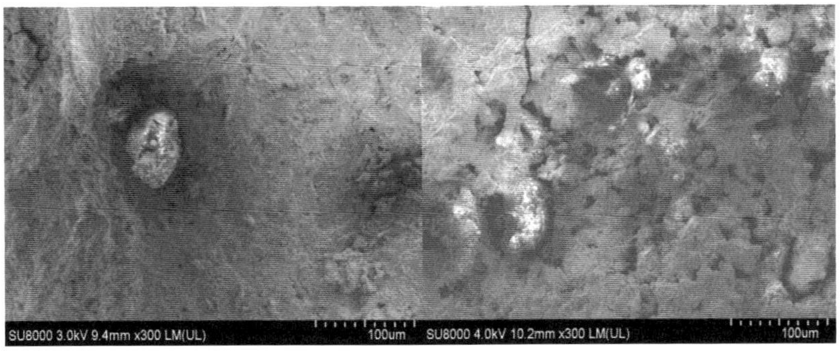

Figure 8. Micro morphology of PEEK cement stone

The micro morphology of cement stone in blank group was observed by scanning electron microscope. It can be seen from Figure 7 that there are many kinds of micro voids in cement stone in blank group. The main components of grade g cement are tricalcium silicate, tetracalcium ferroaluminate and dicalcium silicate. The material itself presents a multilateral pore structure, and the cemented cement stone is easy to present a porous structure. These voids cause the inherent strength and toughness deficiency of cement paste, which needs the supplement of external materials to achieve the effect of strengthening and toughening.

It can be seen from Figure 8 that peek can fill in the cement paste and hinder the crack propagation. When peek is added into cement paste, part of PEEK is filled in the primary void of cement paste, which forms a whole with the original space grid structure of cement paste under the cementation of cement. When the external load does not exceed the yield strength of the cement paste space grid structure, peek, as the filling particles in the cement paste space grid structure, acts as the transmission medium of the external load, deforms itself, absorbs part of the energy, and slows down the force of the external load on the cement paste, so as to improve the ability of the cement paste to resist the external load. On the other hand, when the external load exceeds the yield strength of the space grid structure, the cracks in the cement paste will expand along a certain direction until the whole cement paste is destroyed. However, when the crack propagates to peek, it is necessary to destroy peek or change the propagation direction to bypass the filling material. Peek is a kind of material with high strength and high elastic modulus. It is difficult to destroy peek and continue to expand along the original direction. Most of the cracks bypass the filling material to change the direction of crack propagation and continue to expand along the new direction. This will greatly delay the destruction process of cement stone, reduce the brittleness of cement stone, and enhance the ability of cement stone to resist external load to a certain extent.

4. Conclusions

(1) Peek has little effect on the thickening time and water loss of cement slurry, and the cement slurry has good settlement stability. However, peek will lead to poor rheological properties of cement slurry. In practical application, if the dosage of PEEK is increased, the rheological property of cement slurry should be monitored in time to prevent the cement slurry from being difficult to pump into the well due to too high rheological reading.

(2) The results show that peek can enhance the ability of cement stone to resist external load, and the best dosage is 5%. After curing for 28 days, the flexural strength and impact strength of PEEK cement stone with 5% peek are increased by 31% and 24% respectively compared with those without peek.

(3) The results show that peek can reduce the elastic modulus of cement paste. The elastic modulus of cement paste containing 5% peek is about 50% lower than that of the blank sample, reaching 5.1 GPa, and the elastic toughness of cement paste is obviously improved.

(4) When the cement paste added with PEEK is subjected to external load, peek itself will deform and absorb energy. Meanwhile, peek will prevent and change the crack expansion of cement paste and improve the ability of cement paste to resist external load.

5. Acknowledgments

This work was supported by the National Natural Science Foundation of China (Grant No. 519040334, 5180404-4). And the National Major Science and Technology Projects of China (Grant number 2016ZX05060 and 2017ZX05032004-004).

6. References

[1] H Dong, Y Xiao, *Journal of China University of Petroleum*. 37 (2017)108-113.
[2] S Jiyun, S Lin, L Yong, Natural Gas Industyr. 37 (2017)98-108.
[3] K Agapious, R Gamwell, ARMA. 12 (2016)16-32.
[4] L Zaoyuan, G Xiaoyang, Petroleum drilling technology. 5 (2016)86-92.
[5] L Zaoyuan, G Xiaoyang, Petroleum drilling technology. 18 (2018)52-55.
[6] P Shuo,W Qin, Z Dafu,W Dongmei, Portland Bulletin. 38 (2019)2622-2630.
[7] K. Nova, P. Novak, F. Prusa, J. Kopecek and J. Cech, Metals 9, (2019), p.1.
[8] L Yuliang, L Zaoyuan, X Yuantao, Z Biao, S Hao, S Donghua, S Jinfei, Portland Bulletin.38 (2019)339-343.
[9] F Cuzzocrea, A Ivone, E Jannelli, A Fioruzzi, Musculoskeletal Surgery. 103 (2019)62-71.
[10] Z Xiaoiqng, Z Yifang, *Guangxi Medical Journal*. 41 (2019)2144-2146.
[11] L Yongwei, Chemical Engineer. 33 (2019)61-65.
[12] C Xi, Y Haiyang, International Journal of Stomatology. 45 (2018)657-665.
[13] E. Wielgosz, T. Kargul, J. Therm. Anal. Calorim. 119 (2015) 1547-1553.
[14] M. Ramezani, T. Neitzert, J. Achieve. Mater. Manuf. Eng.55/2 (2018), p.790

IOP Conf. Series: Materials Science and Engineering 1133 (2021) 012007 doi:10.1088/1757-899X/1133/1/012007

Analysis on the hot deformation flow stress curves of novel 6082 aluminium alloys with Mn addition

Yong Li [1, 2], Xiaoming Qian [1a, *], Yu Zhang [1], Zhaodong Wang [1,*], Jiadong Li [1]

[1]State Key Laboratory of Rolling and Automation, Northeastern University, Shenyang, 110819, China
[2]Guangxi Advanced Aluminum Processing Innovation Center Co. LTD, Nanning, 530007, China
[a]qxmwww@126.com

Abstract. Two types of 6082 alloy, base alloy free of Mn and novel 0.75Mn alloy with an addition of 0.75% Mn, were prepared. A heat treatment of 450 °C for 6 h was performed, during which large numbers of α-Al (Fe, Mn) Si dispersoids precipitated in 0.75Mn alloy. Uniaxial compression tests were carried out and a series of flow stress curves were obtained at temperature range of 400–550 °C and at strain rate range of 0.001s-1-1s-1. The results revealed an impressive difference between the flow stress curves of base and 0.75Mn alloy that after the yield plateau, the flow stresses of base alloy remained fairly steady at certain levels while the flow stress in 0.75Mn alloy decreased along with strain until the end of deformation. The reason for the decrease of flow stress in 0.75Mn alloy was neither the dissolution of dispersoids nor the dynamic recrystallization. The decrease of flow stress was attributed to the loss of Mn solid solution strengthening effect based on the Mn diffusion rate-controlling mechanism and "pipe diffusion".

1. Introduction

The main strengthening mechanism of traditional 6082 aluminium alloys at room temperatures is based on the fine nano-scale Mg^2Si precursors [1], [2]. However, the coarsening and/or dissolution of Mg^2Si starts to take place at the temperature exceeding 200 °C (similar as over aging), which leads to severe deterioration of mechanical properties [3].

To develop novel alloys that can be applied at elevated temperatures, our previous study explored the way using α-Al (Mn, Fe) Si dispersoids to strengthen 6082 alloy at elevated temperatures [4]. The results show that large number of dispersoids was induced by adding Mn element and through a low-temperature homogenization of 450 °C for 6 h. The dispersoids-strengthened alloy yielded significantly increased high-temperature flow stresses. An increase in hot deformation activation energy from 191 kJ/mol for the Mn-free alloy to 301 kJ/mol for the alloy with 0.75% of Mn was derived from a series of hot deformation flow stress curves.

The feature of flow stress curves are important indicators to the changes in microstructures and can reflect the balance between dynamic hardening and dynamic softening during the hot deformation process [5], [6]. In the typical hot deformation flow stress curves, the flow stress usually increases sharply at the early stage of deformation until reaching a peak plateau [7], [8]. After then, flow stress curves features can be generally divided into several typical cases: i) increase, ii) stay stable and iii) decrease along with increasing strain. The different trends of flow stresses after the peak plateau are closely related the hot deformation conditions [9], [10] and microstructural evolutions.

For the first case of the continuous increase of flow stress along with strain, it is usually associated with the condition when the work hardening overcome the effect of dynamic softening. Shakiba [11] studied the flow stress behavior of a 1xxx dilute aluminium alloy, the resulted shows that continuous increased flow stresses after yield plateau were observed at the low temperature and high strain rate deformation temperatures of 350°C/1s^{-1} when dynamic recovery (DRV) was the predominant softening mechanism. While with the increased deformation temperature by 400°C at the fixed strain rate of 1s^{-1} or at a lowered strain rate of 0.001s^{-1} at fixed deformation temperature of 350°C, the feature of flow stress curves switched to the second case that flow stress kept steady at some certain levels along with strain, which indicated the achieved dynamic equilibrium between work hardening and softening [8].
The third case that flow stress after yield plateau declines along with strain is associated with the fact that the rate of dynamic softening is enhanced and are already exceed the rate of dynamic hardening. The enhanced dynamic softening could be resulted from thermal softening [12]-[14], dynamic recovery (DRV) [15], dynamic recrystallization (DRX) [16]-[18] and coalescence of precipitates [19], [20].
Thermal softening usually occurred when the deformation was conduct at high strain rate [12]-[14]. Shi reported the thermal softening phenomena in 7150 aluminium alloys [14]. When a 10s^{-1} strain rate was set, the deformation was finished within only 0.08 second. Such a short deformation time resulted that the heat produced from deformation could not fully released. Hence an as high as 35 °C temperature rise was induced in the tested sample, led to a thermal softening mechanism during hot deformation.
DRX in the microstructures was an important softening mechanism during hot deformation. DRX could often be triggered when deformation was performed at high temperatures and low strain rates [14]. During deformation, dislocations and subgrains were progressively developed inside the deformed grains, the slow straining process and high temperature allows the accumulated substructures progressively converted into grain boundaries via the combining of dislocations and rotation of subgrains [5], [21]. The newly formed grains are featured as free of substructures and are much softer than the original deformed ones, leading to the dynamic softening during hot deformation. [22], [23].
The softening mechanism of coalescence of precipitates involves the loss of strengthening effect exerted by the coarsen or the dissolution of fine precipitated particles [19]. During hot deformation, once coalescence of precipitates occurred, the strong pinning effect of fine precipitates on dislocation movement was severely weakened, leading to enhanced rate of dynamic softening. Besides, the coarsening and/or dissolution of precipitates further improves the rate of DRV/DRX, which resulted in a more remarkable dynamic softening [24], [25].
The analysis of flow stress curves is a very classic topic in the field of aluminium research. A series of investigation, including microstructural observation, data-based analysis, modeling and predicting, have been carried out by researchers [26]-[29]. The investigation specifically on the novel dispersoids-strengthened 6082 alloys still cannot found in the open literatures. The mechanism behind the changes in flow stresses of novel dispersoids-strengthened 6082 alloys need to be better understood.
In the present study, flow stress curves of base 6082 alloy free of Mn and a novel dispersoids-strengthened 6082 alloy were obtained by performing hot-compression tests with temperatures ranged from 400 to 550 °C and strain rates from 1 to 0.001 s^{-1}. The hot deformation flow stress curves were analyzed, typical microstructural evolution during deformation was examined. The focus of this study is to understand the main mechanism of flow stress behavior in the novel dispersoids-strengthened 6082 alloy.

2. Experimental methods

Two 6082 type alloys, base alloy free of Mn and 0.75Mn alloy with a 0.75% Mn addition, were used for the investigation. The chemical compositions are shown in Table 1. The billets were cast prepared using direct chill cast with 101 mm diameter. Heat treatment of 450 °C for 6 h with a 100°C/h heating rate was performed in order to promote the precipitation of dispersoids in the aluminium matrix [4]. Cylindrical specimens with diameter 10 mm and length 15 mm were prepared for uniaxial hot-compression tests. The hot-compression were performed on a Gleeble 3800 thermo-mechanical testing unit. The specimens were heated at 2 °C/s and held to the desired temperatures for 180 s to ensure a homogeneous temperature distribution. The specimens were then compressed to a total true strain of 0.75 with the temperatures ranged from 400 to 550 °C and strain rates from 1 to 0.001 s^{-1}. The flow stress date was obtained by the testing unit and then saved automatically. A series of flow stress curves were drawn in function of true strain and true stress at different deformation temperatures and strain rates. For the observation of microstructures, an optical microscope (Nikon, Eclipse ME600), a scanning electron microscope (SEM, JEOL-6480LV) and a transmission electron microscope (TEM, JEM–2100) operating at 200 kV were applied. An electron back-scattered diffraction (EBSD) analysis was used to check the grain structures. All Euler orientation maps were applied in the EBSD image analysis. Moreover, a quantitative analysis of the dispersoid particle was performed based on the SEM images of the polished sample surfaces, some of the samples were etched with a 0.5% HF solution for 40 s to further reveal the detail of microstructures. A Sigmascope SMP10 eddy-current device was used at room temperature to measure the electrical conductivity to follow the solid solution levels.

Table 1. Chemical compositions (wt.%) of the experimental alloys.

Alloys	Mg	Si	Fe	Mn	Al
Base	0.79	1	0.18	-	Bal.
0.75Mn	0.84	1.02	0.23	0.72	Bal.

3. Results

3.1. Hot deformation flow stresses

Hot deformation was conduct at temperatures ranged from 400 to 550 °C and strain rates from 1 to 0.001 s^{-1}. Figure. 1 shows a series of deformation true stress–true strain curves obtained under the given experimental conditions. For all the deformation conditions, the flow stress raised sharply at the beginning of the deformation until reaching peak stresses. Subsequently, 3 types of trend in flow stress curves, continuous but slow increase, remaining stable at certain levels and decrease smoothly with increasing strain, were observed, depending on the different alloys and deformation parameters.

The continuous increase of flow stresses after yield were mainly presented in the base alloy when the strain rate was 1s^{-1} at all four deformation temperatures as well as in the 0.75Mn alloy at 1s^{-1} strain rate under the 400 and 450 °C deformation temperatures. For the case of flow stress remained stable at a certain level after peak, the typical conditions were presented in the base alloy when the deformation strain rates of were 0.1, 0.01 and 0.001s^{-1} at all four deformation temperatures (Figure. 1a, c, e). The decrease of flow stress after peak with increasing strain could be only distinguished in the 0.75Mn alloy, the corresponding conditions covered nearly all the deformation regimes except that of the 1s^{-1} strain rate at 400 and 450°C (Figure. 1b, d, f).

Besides, some evident general tendencies were observed, i.e., the level of flow stress decreased with the increase in deformation temperatures and increased with the increase in strain rates for both base and 0.75Mn alloys. Additionally, the overall flow stress levels in the 0.75Mn alloy were higher than those of base alloy under same deformation parameters.

To further demonstrate the flow stress of two experimental alloys, peak flow stresses in both alloy and flow stress at 0.75 strain in 0.75Mn alloy were chosen as the typical values to clarify the features of flow stress curves. As shown in Figure. 2, in general, overall higher peak flow stresses in 0.75Mn alloy (Figure. 2b) than those of base alloy (Figure. 2a) were distinguished. The peak flow stress increased with increasing strain rate and decreasing temperature. Flow stress at 0.75 strain in 0.75Mn alloy showed the similar tendencies as those of peak flow stress, while an extra feature was that the decrease of stress with decreasing strain rate was more remarkable than those of peak stress (Figure. 2b).

In order to further characterize the differences in the peak flow stress and flow stress at 0.75 strain in 0.75Mn alloy, a decline ratio of flow stress, R_d, is introduced and defined as Eq. (1) [7]:

$$R_d = \frac{\sigma_p - \sigma_{0.75}}{\sigma_p} \times 100\% \qquad \text{Eq. (1)}$$

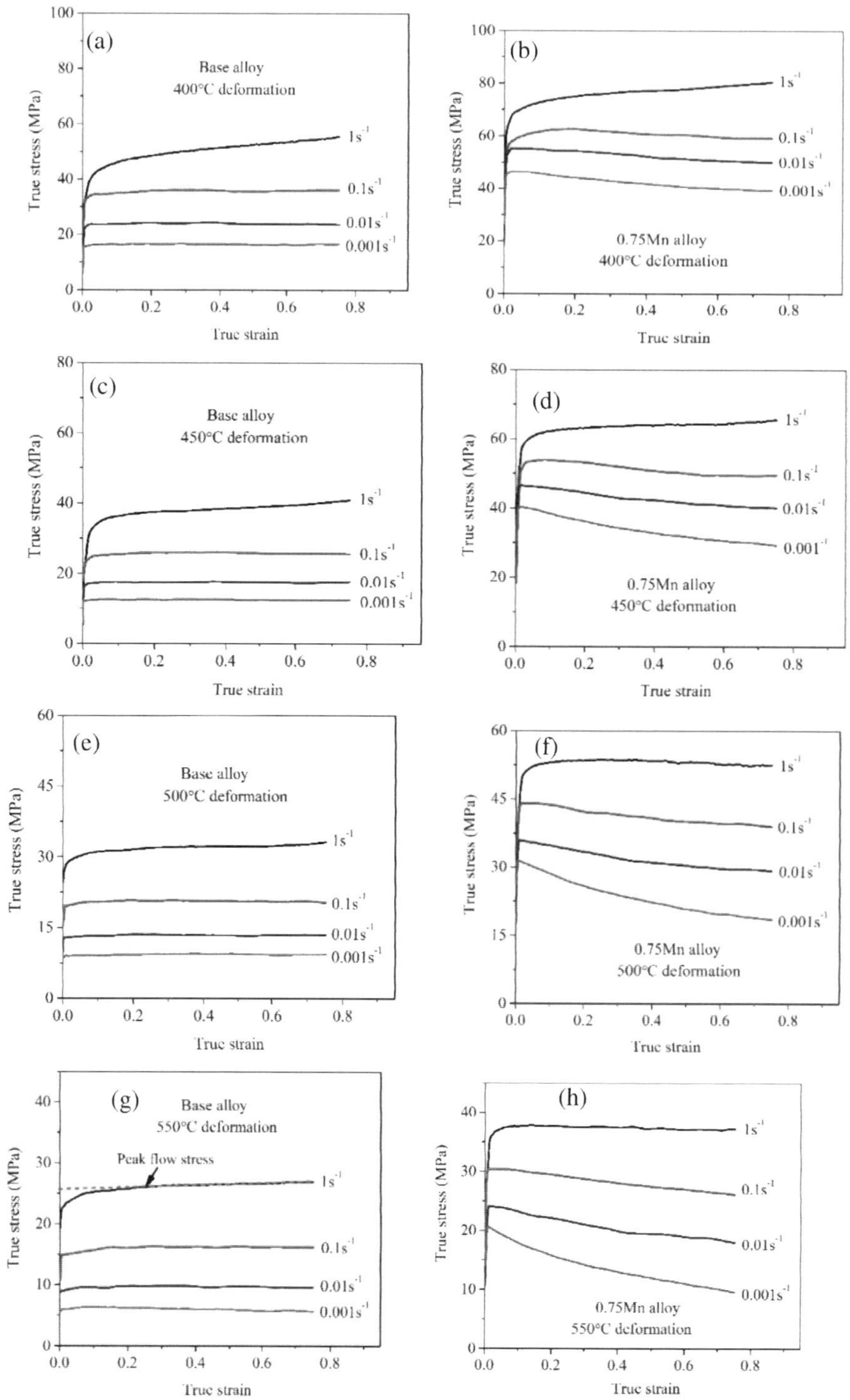

Figure 1. True stress–true strain curves of hot compression deformation, (a), (c), (e) and (g) base alloy; (b), (d), (f) and (h) 0.75Mn alloy.

IOP Conf. Series: Materials Science and Engineering 1133 (2021) 012007 doi:10.1088/1757-899X/1133/1/012007

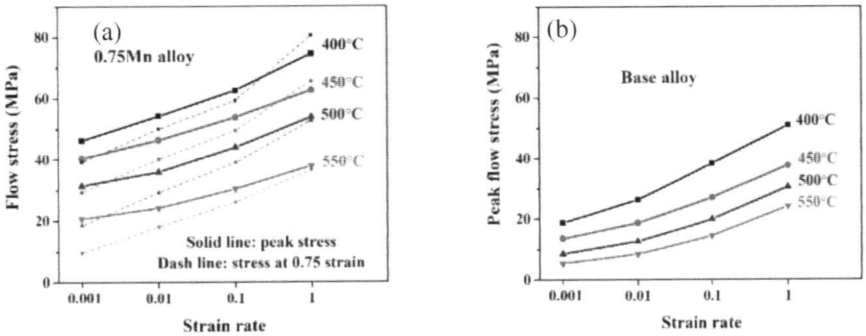

Figure 2. Peak flow stresses in base alloy (a), peak flow stress and stress at 0.75 strain of 0.75Mn alloy (b) at different hot deformation temperature as a function of the strain rate.

Where σ_p is the value of peak stress and $\sigma_{0.75}$ is the value of flow stress at the 0.75 strain of the deformation. Figure 3 shows the results of R_d of 0.75Mn alloy in function of strain rate at different deformation temperatures. In general, R_d decreased with increasing strain rate and decreasing temperature. Specifically, when the deformation strain rate was $1s^{-1}$, -8.3% at and -4.3 of R_d value were derived from 400°C and 450°C deformation data respectively. These negative values respected the increase of flow stress along with strain after yield plateau (Figure. 1). As the deformation temperature increased by 500 and 550°C at the $1s^{-1}$ strain rate, R_d became greater than 1, corresponded with the drop tendency of flow stress curve after yield plateau at these temperatures. Regarding the strain rates of 0.1, 0.01, $0.001s^{-1}$, R_d values for all four deformation temperatures became positive value, which was associated with the increased flow stress curves in all four deformation temperatures.

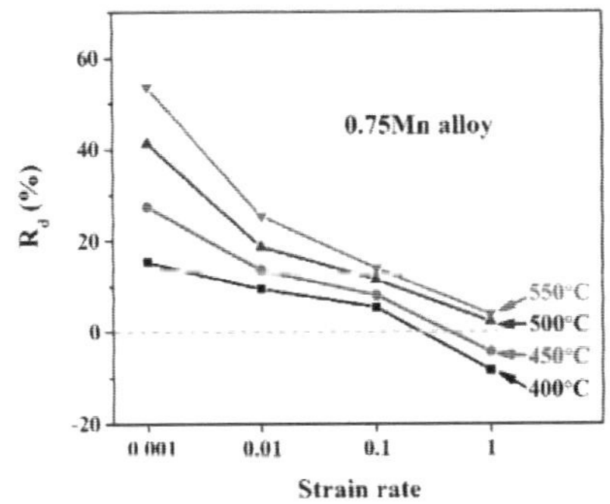

Figure 3. Decline ratio of flow stress of 0.75Mn alloy in function of strain rate at different deformation temperatures.

3.2. Microstructures evolution during deformation

The experimental 0.75Mn alloy was designed as a dispersoid-strengthening alloy. Addition of 0.75% Mn and the heat treatment ensured the fully precipitation of dispersoids to promote the high temperature mechanical properties [4], base alloy free of Mn was prepared to make a comparison. Figure. 4 shows the microstructures of the two alloys after deformation of 450 °C/ 0.1 s^{-1}. Mainly elongated grains were observed in the samples after hot deformation. Fe-rich intermetallics (white in the inset SEM images) and primary $Mg^{2}Si$ particles (black) were distributed along the dendrite boundaries (Figure. 4a, b). On the surface of etched samples, the dispersoids can be clearly revealed in 0.75Mn alloy (Figure. 2b, c), these dispersoids were produced through the Mn addition and the 450°C/6h heat treatment. Our previous study had confirmed these dispersoids as α-Al (Fe, Mn) Si phase [4]. SEM image (Figure. 4c) shows that the dispersoids locate at the region inside the dendrite cells. Meanwhile, dispersoids free zone formed near the dendrite boundary regions where

IOP Conf. Series: Materials Science and Engineering 1133 (2021) 012007 doi:10.1088/1757-899X/1133/1/012007

intermetallics was observed. The morphology of dispersoids was closely observed in TEM image (Figure. 4d), they appeared mostly in round or elliptical shapes in the two-dimensional image. Besides, dislocations interacted with the dispersoids were observed, indicating a retarding effect of dispersoids on dislocation motion during deformation.

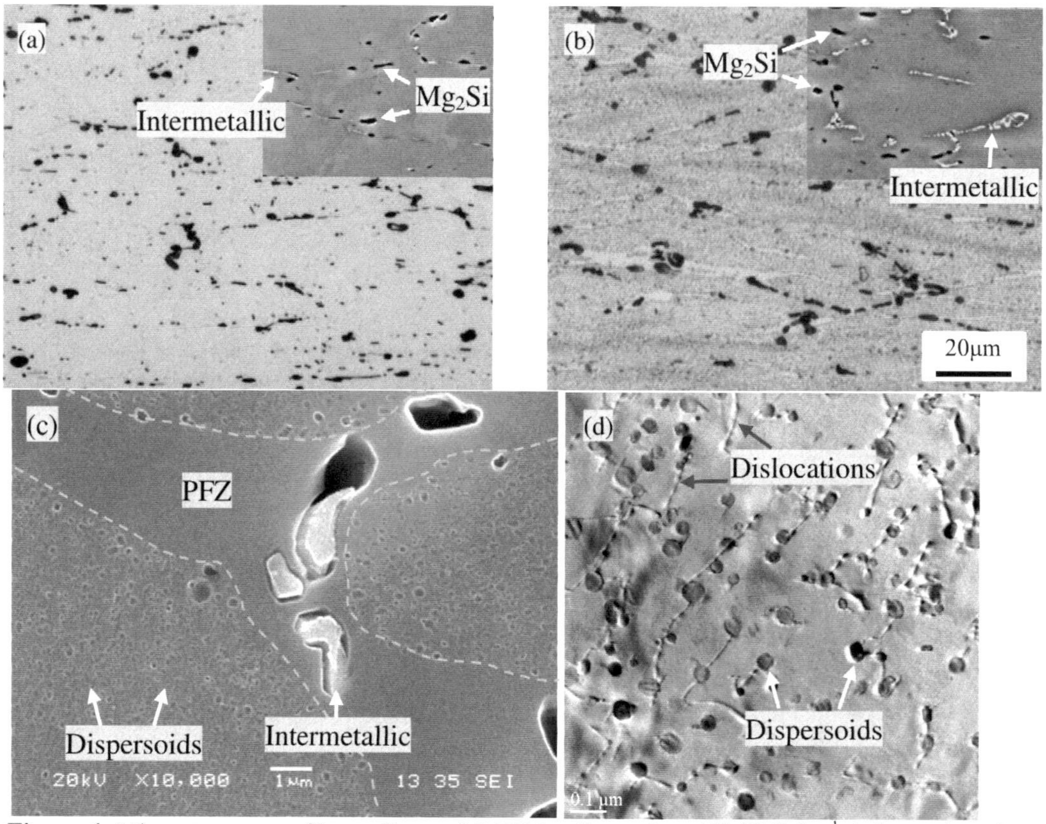

Figure 4. Microstructures of base alloy and 0.75Mn alloys after 450 °C/0.1 s^{-1} deformation: SEM images of etched surface (with non-etched image inset) of (a) base alloy and (b) 0.75Mn alloy, enlarged SEM image of etched surface of 0.75Mn alloy (c), (d)bright field TEM image of 0.75Mn alloy.

The dispersoids number density before and after deformation were quantified base on the SEM images. The results suggested that 0.75Mn alloy had a dispersoids number density of 11.5μm^{-2} before deformation. After deformation was performed, the change of dispersoids number density was fairly limited at most of deformation conditions, while only after 550°C/0.001s^{-1} deformation, an obvious decrease in the population of dispersoids was observed. Figure. 5 shows the typical etched surface of 0.75Mn alloy after 450°C/0.1 s^{-1} and 550°C/0.001s^{-1} deformation. In the Figure. 5a, the dispersoids number density is 11.5μm^{-2}, which is close to that of before deformed samples, while in Figure. 5b a much small population of dispersoids (dispersoids number density dropped by 5.6 μm^{-2}) was observed, suggesting the dissolution of dispersoids took place during the high temperature-low strain rate deformation.

Evolution of primary Mg_2Si particles were followed. At the low deformation temperatures range of 400-500°C, no occurrence of Mg_2Si dissolution was observed for all four strain rates. At 550°C of deformation temperature, strain rates no less than 0.01s^{-1} still was still not able to make full dissolution of Mg_2Si (Figure. 6a, c) in both base and 0.75Mn alloys. Only the lowest strain rate of 0.001s^{-1} with highest deformation temperature of 550°C could lead the full dissolution of Mg_2Si (Figure. 6b, d).

Grain structures were investigated using EBSD technology. Figure. 7 shows All Euler orientation maps of two alloys under the three typical deformation conditions. In the orientation maps, different colors represent the orientation differences of grains and subgrains. Elongated grains perpendicular to the compression direction were observed in all deformed samples. At the 450 °C/0.1 s^{-1} deformation condition, a large amount of white and green lines was observed in the elongated grains, implying the high densities of dislocation cells and subgrains (Figure. 7a, d) in both base and 0.75Mn alloys. These EBSD map were typically characterized as dynamic recovery (DRV) [5]. When the deformation temperature raised to 550°C at $0.1s^{-1}$ strain rate, the density of dislocation cells and subgrains were obvious lowered (Figure. 7b, e), the reduction of substructures in base alloy was much evident than that in 0.75Mn alloy. Besides, some grains nearly free of internal substructures were observed in base alloy (arrows in Figure. 7b), indicating occurrence of dynamic recrystallization (DRX) [5], [21], [22]. However, DRX was not observed in 0.75Mn alloy, DRV was still the predominant mechanism after deformation. When the strain rate decreased by $0.001s^{-1}$ at 550°C deformation temperature, a nearly full DRX took place in base alloy, reflected by the almost disappearance of substructures (Figure. 7c). In 0.75Mn alloy, some small grains free of internal substructures were observed near the original grain boundaries (arrows in Figure. 7f), suggesting the occurrence of partial DRX during hot deformation.

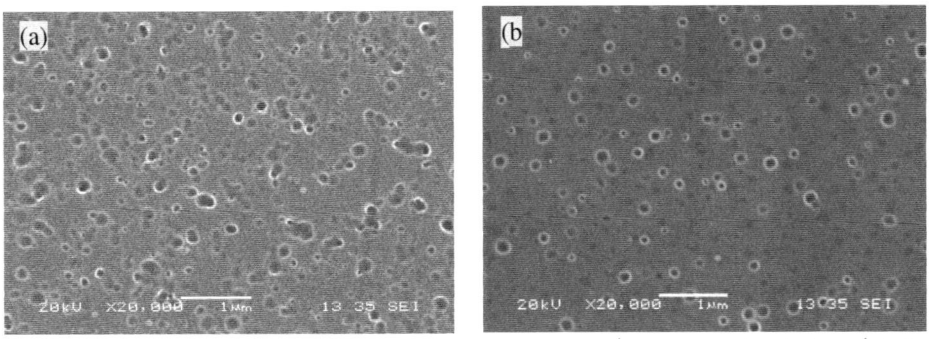

Figure 5. Dispersoids in 0.75Mn alloy after (a) 450°C/0.1 s^{-1} and (b) 550°C/$0.001s^{-1}$ deformation.

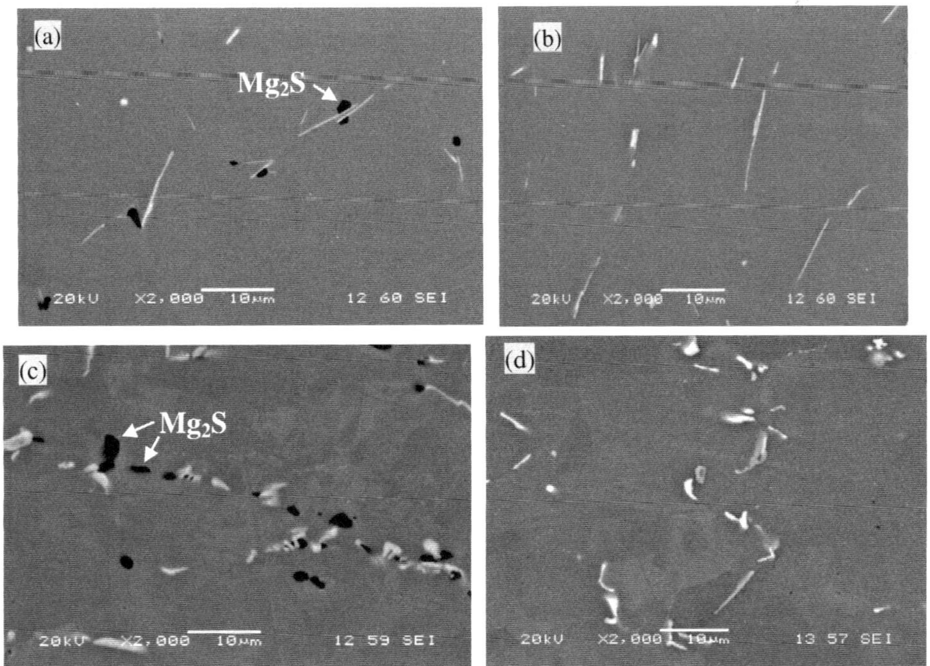

Figure 6. SEM image after deformed at 550°C with strain rate of (a) 0.01 s^{-1} in base alloy, (b) 0.01 s^{-1} in 0.75Mn alloy, (c) 0.001 s^{-1}in base alloy and (d) 0.001 s^{-1} in 0.75Mn alloy.

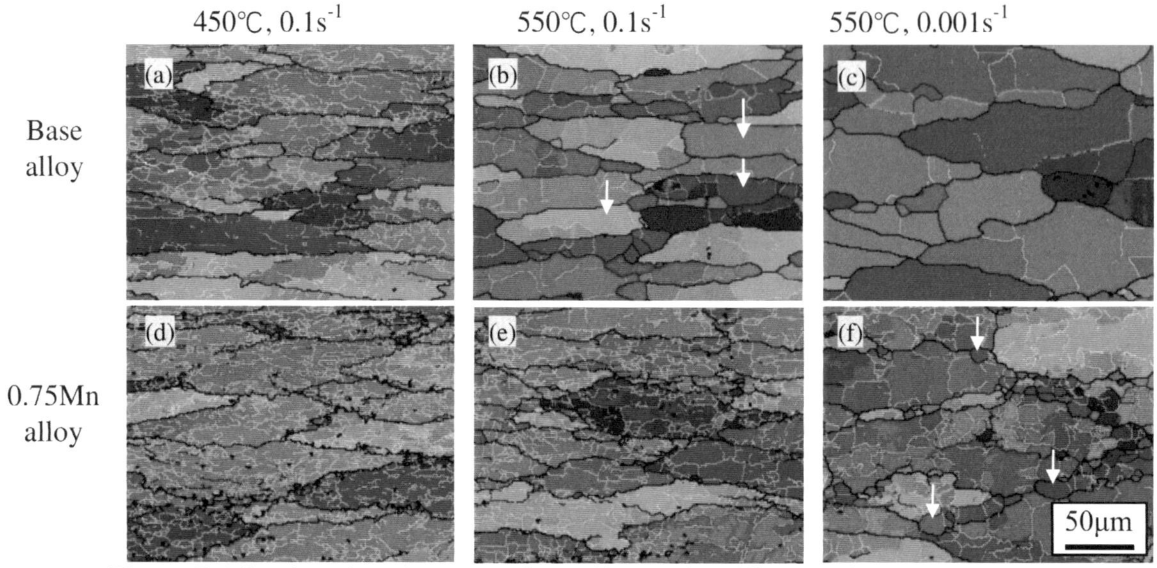

Figure 7. All Euler orientation maps of two alloys under different deformation conditions. The boundary misorientation angles are marked by white lines 2-5°, green lines 5-15° and black lines > 15°. A step size of 0.5 μm was set during the orientation map scanning.

4. Discussion

Hot deformation flow stress and the microstructures evolution were investigated based on 6082 base alloy and a novel dispersoids-strengthened 0.75Mn 6082 alloy. The hot deformation flow stress curves exhibited various features, which was closely related to deformation parameters and the evolution of microstructures reflecting the dynamic balance between hardening and softening.

4.1. Increased flow stress after yield plateau

The continuous increase of flow stresses after yield plateau were mainly presented in the base alloy when the strain rate was $1s^{-1}$ at all four deformation temperatures as well as in the 0.75Mn alloy at $1s^{-1}$ strain rate under the 400 and 450 °C deformation temperatures [Figure. 1 and 3]. During deformation process, dislocations were introduced progressively. With increasing strain, the accumulation and pile-up of dislocation became more severe, resulting in the high rate of dynamic work hardening [30]. Meanwhile, the occurrence of dynamic softening via the combining and rearrangement of dislocations and subgrains, which was associated with DRV, could partially release the accumulated energy and lead to a weakening of work hardening [31]. In case of high strain rate of $1s^{-1}$, the while duration of deformation was only 0.8 second, in such a short time, DRV was not able to take place in highly active degree, corresponding dynamic softening rate stayed at a lower level than that of work hardening and was responsible for the continuous increasing flow stress along with strain.

4.2. Steady flow stress level after yield plateau

Steady flow stress level after yield plateau was observed main in the base alloy when the deformation strain rates of were 0.1, 0.01 and $0.001s^{-1}$ at all four deformation temperatures [Figure. 1]. The steady flow stress implied that an equilibrium balance between dynamic hardening and dynamic softening was reached and was maintained until the end of deformation. When the deformation was performed under lower strain rate, the longer duration allows the high degree of DRV and DRX take place (Figure. 7b, c), resulting in the enhanced dynamic softening. when an equilibrium balance between dynamic hardening and dynamic softening was reached, the flow stress curves exhibited the steady feature. The similar steady flow stress was also observed in the dilute aluminium alloy [32] and some Al-Mg-Si alloys [27], [33], [34].

4.3. Decreased flow stress after yield

The decrease of flow stress after yield plateau with increasing strain was presented only in the 0.75Mn alloy. The corresponding conditions covered all the deformation regimes except that of the $1s^{-1}$ strain rate at 450 and 500°C.

Decline of flow stress after yield plateau was a sign that the dynamic softening rate had exceed that of dynamic hardening. It was reported in literatures that the dynamic softening could be associated with coalescence of precipitates [19], [20], dynamic recrystallization (DRX) [12]-[14] and thermal softening [12]-[14].

In 0.75Mn alloy, the overall higher flow stress level than those in base alloy could be the result of the strong strengthening effect of α-Al (Fe, Mn)Si dispersoids (Figure. 1, 2) [4]. Dispersoids number density was an important parameter influencing flow stress levels and contributing the change in flow stress curves. it is reported that [35] strengthening of α-Al(Fe, Mn)Si dispersoid is the predominant strengthening mechanism for a Al–Mn–Mg–Si alloys. In this study, the decline of flow stress after yield can be interpreted as being due to the dissolution of dispersoids. The dispersoids number density before deformation in 0.75Mn alloy was $11.5\mu m^{-2}$ (Figure. 4a), whereas after deformation at 550°C with $0.001s^{-1}$ strain rate, the dispersoids number density dropped by $5.6\mu m^{-2}$ (Figure. 4b), indicating that a loss of 49% population of dispersoids during the deformation, which may be responsible for the severe drop in flow stress from 20.7MPa at peak by 9.6MPa at 0.75 strain. However, for all of the rest condition of $0.1s^{-1}$, $0.01s^{-1}$ and $0.001s^{-1}$ strain rate at all four deformation temperatures, where no sever dissolution of dispersoids took place, still it could be observed the decline of flow stress after yield (Figure. 1 and Figure. 3), which implying a fact that the dissolution of dispersoids was not the direct reason responsible for the declining flow stress.

In 0.75Mn alloy, DRX was observed in some deformed samples, for example, $550/0.001s^{-1}$ in 0.75Mn alloy. It is reported that, in Al–Mg–Si–Cu aluminium alloy, when the DRX occurred, an obvious decline or fluctuation in flow stress was observed due to the releasing of stored strain energy via formation of new grains [26, 30]. Wherever, in 0.75Mn alloy no evident flow stress fluctuation along with strain could be observed, all the flow stress curves exhibited quite smooth decline rates. Besides, under the deformation conditions where no DRX occurred (for example, $550°C/0.1s^{-1}$ condition), still the decline of flow stress after peak plateau was observed, indicating that DRX may be not the key factor determining the decline of flow stress.

Thermal softening was reported to be a softening mechanism under the conditions when the deformation was conduct at high strain rate [12]-[14]. In 7150 aluminium alloys [14], When train rate was set as $10s^{-1}$ s, a 35 °C temperature rise was induced in tested samples due to the strain heat could not fully released within as short as only 0.08 second of deformation duration. This raised temperature resulted in a thermal softening mechanism during hot deformation. However, in the present study, the highest strain rate is $1s^{-1}$, no evident temperature raise in deforming specimen was tested, which ruled out the possibility of thermal softening.

The above mechanisms of dispersoids dissolution, DRX mechanism and thermal softening seem to be difficult to clearly explain the decline of flow stress after yield plateau in 0.75Mn alloy. A solution diffusion rate-controlling mechanism that involves the changes in the solid solution strengthening may give an explanation for the decline of flow stress phenomenon in 0.75Mn alloy.

On view of solutions strengthening in the Al matrix, changes in solution levels can significantly influence the hot deformation flow stress of aluminium alloys [36], [37]. Shakiba studied the hot workability of dilute Al-Fe-Si alloys and found that increasing the iron solution level from 0.1 to 0.7% produced 11 to 32% increase in the hot deformation flow stress in a dilute Al-Fe-0.1Si alloy due to the solid solution strengthening effect of Si and Fe elements [11]. The increased flow stress level resulted from solid solution level was also reported in the 1xxx alloy during homogenization [32]. A rise in the homogenization temperature from 550 to 630 °C resulted in the increased flow stress by 15–45% in a 1xxx alloy due to the enhanced release of solution element to the matrix at high homogenization temperatures.

In the present study, Mg, Si, Fe and Mn was involved in view of solid solutions. Technically, it is not an easy work to directly test the solid solution levels of element in the aluminium matrix. In this case, electrical conductivity is usually used as an indicator to follow the change in solid solution level [32]. In this study, the electrical conductivity of base and 0.75Mn alloy after heat treatment (before deformation) were 48.9%IACS and 47.4%IACS respectively. This lower electrical conductivity in 0.75Mn alloy implied that, despite a 450°C/6h heat treatment was conduct to promote the precipitation of Mn-containing dispersoids, there still existed a certain amount of Mn remained in Al matrix. This extra Mn solute present in 0.75Mn alloy could pose a significant solid solution strengthening effect during hot deformation.

The impact of different types of solute elements on hot deformation flow stress behavior was studied by Shakiba et. al. It is claimed that the solute diffusion rate in aluminium alloys was the key factor determining the flow stress levels of hot deformation. When different types of solution elements were individually added in the aluminium alloys, the hot deformation flow stress level is inversely proportional to the diffusion coefficient of the added solute elements, which is referred as solution diffusion rate-controlling mechanism [38].

Figure. 8 shows the diffusion coefficient of Mg, Si, Fe and Mn in aluminium as well as that of Al self-diffusion [39]. Mg and Si have the slightly higher diffusion coefficients in Al than the Al-self diffusion, while Fe have moderately lower diffusion coefficients than those of Al-self diffusion. Mn exhibited the lowest diffusion coefficients at the given temperature range. According to the rate-controlling mechanism, Mn, which have lowest diffusion coefficients, should has the much stronger impact on the flow stress than those of Fe, Si and Mg.

For the case of Mn solute in 0.75Mn alloy, Mn solute atoms had a strong solid solution effect because of its low diffusion rate in aluminium according to the diffusion rate-controlling mechanism. At the early stage of deformation, as soon as the elastic deformation was started, the dislocations begin to produced. In this case, Mn solute atoms act as the barrier for the dislocation motion, which resulted in a high deformation resistance and lead to a hardening effect [38]. As the strain goes on, until the work hardening and dynamic softening meet at a balance rate, the yield point was reached.

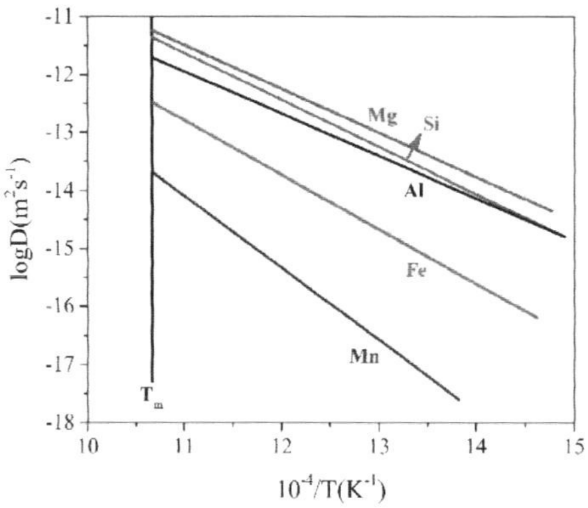

Figure 8. Diffusion coefficients of Mg, Si, Fe and Mn in aluminium.

After peaks, flow stress begins to decline after peak in 0.75Mn alloy, which suggested that the solution strengthening may be influenced by other factors. A "pipe diffusion" mechanism may explain the change in the solution strengthening along with strain. "Pipe diffusion" refers a phenomenon that when carried by dislocations, the solute atoms diffusion switched from the static diffusion to dynamic diffusion mechanism, so that the diffusion rate increased remarkably. Dislocations served as fast diffusion paths for solute atoms diffusion in the matrix, thereby permitting rapid dynamic diffusion of solution atoms [40], [41]. In 0.75Mn alloy, the dislocations were produced dynamically in a considerable high rate along with the strain. When the dislocations encountered Mn atoms, a "pipe diffusion" mechanism was turned on. It is reported that after the "pipe diffusion" take place, the diffusion rate became as high as several order as that at static diffuse condition [41]. In case of significantly increased Mn atom diffusion rate, the deformation resistance was weakened according to the diffusion rate-controlling mechanism. When the "pipe diffusion" take place, Mn solute atoms somehow could not act as the barrier of dislocation motion any longer. Furthermore, when the unpinned dislocations glided passing the static Mn atoms in the matrix, static Mn atoms were "wiped out" by the dislocations and start the "pipe diffusion", correspondingly resulting in the further severe loss of strength of the alloy. Therefore, the diffusion rate-controlling mechanism and the "pipe diffusion" seems to give a reasonable explanation for the decline of the flow stress in 0.75Mn alloy.

In base alloy free of Mn, the steady flow stress after yield could also be explained by the diffusion rate-controlling mechanism and "pipe diffusion". According to the diffusion rate-controlling mechanism, the diffusion coefficients of Mg and Si are much higher than those of Mn and was very closed to those of Al-self diffusion. This suggested that the impacts of Mg and Si in solution were much weaker than those of Mn and even approximately equaled to the Al matrix. After yield plateau, when the "pipe diffusion" was highly active, the impact of strength driven from Mg and Si solution strengthening was further weakened. Hence, no evident loss in strengthen could be produced after the turn on of "pipe diffusion" in Mg and Si, explaining the fairly steady flow stress along with strain until the end of deformation. Similar steady flow stress feature along with strain was also seen in the Al-Mg-Si alloys in the works of [27], [32], [33], in which the element contents of studied alloys were similar as those in base alloy of this study.

Fe is another element added in both alloys which has a moderately low diffusion coefficient. Theoretically, Fe has lower diffusion coefficients and should have an evident effect on the flow stress. However, the solute ability of Fe in Al matrix was quite limit (about 0.003% at 450°C) [32] after the heat treatment at 450°C for 6h. In the case of such low level of Fe element in the matrix, it is difficult for Fe to make a strong impact on the flow stress.

To sum up, the decline of flow stress after peak in 0.75Mn alloy was due to the significantly increased diffusion rate of Mn solution via "pipe diffusion", which lead to the loss of Mn solid solution strengthening along with strain. The effect of Mg and Si played a limit role on the declined flow stress owing to their high diffusion coefficients. Fe could not provide a strong influence on the flow stress because of its low solubility.

Another phenomenon of the dissolution of primary Mg_2Si was observed in both base and 0.75Mn alloy at only 550°C/0.001s^{-1} deformation condition. It is reported that [42], [43] fully dissolution of primary Mg_2Si in Al-Mg-Si alloys made the Mg and Si atoms were released to the Al Matrix, led to a significant increased flow stress level due to an improved solid solution strengthening. However, in present work when dissolution of primary Mg_2Si was observed under 550°C/0.001s^{-1} deformation condition no evident sing of increase nor even a compensation of the decrease in flow stress after yield along with strain was observed. According to the diffusion rate-controlling mechanism and "pipe diffusion", as soon as the Mg and Si released into the Al matrix, their diffusion was switched to "pipe diffusion" and loss their pinning ability on the dislocations motion immediately. Therefore, the impact of dissolution of primary Mg_2Si on flow stress was not significant.

5. Conclusions

The hot deformation tests were carried out on the base and novel dispersoids-strengthened 0.75Mn 6082 alloys in a temperature range of 400–550 °C and strain rate range of 0.001s^{-1}-1s^{-1}, a series of flow stress curves were obtained. Prior to the hot deformation, a heat treatment of 450 °C for 6 h was conduct, a large numbers of α-Al(Fe, Mn)Si dispersoids were precipitated in 0.75Mn alloy. Conclusions were drawn as followed:

(1) Deformation flow stress levels increased with increasing strain rate and decreasing temperature, 0.75Mn alloy had overall higher flow stress level than those of base alloy.

(2) For base alloy, when deformed at 1s^{-1} strain rate, flow stress curves showed continuous increase after the yield plateau, while at 0.1, 0.01 and 0.001 strain rates flow stress remained at certain levels at all four deformation temperatures of 400, 450, 500 and 550°C.

(3) Regarding the 0.75Mn alloy, the increase of flow stress after yield plateau was observed only under the 1s^{-1} strain rate at 400 and 450°C of deformation temperatures. For the rest of deformation conditions, including 1s^{-1} strain rate at 500 and 550°C and 0.1, 0.01, 0.001s^{-1} strain rates at all four deformation temperatures, the flow stress after yield plateau exhibited the decline tendency.

(4) Under 550°C/0.001s^{-1} deformation condition, full dissolution of primary Mg_2Si occurred in

both alloy, dissolution of dispersoids took place in 0.75Mn alloy. Partial DRX was observed in base alloy at 550°C/0.1s^{-1} deformation condition. At 550°C/0.001s^{-1} deformation condition, nearly full DRX in base alloy and partial DRX in 0.75Mn alloy was recognized.

(5) Neither the dissolution of dispersoids nor the DRX was responsible for the decrease of flow stress in 0.75Mn alloy. The decreased flow stress was explained using a solution diffusion rate-controlling mechanism. When diffusion of Mn solution was active via "pipe diffusion", the diffusion rate of Mn increased significantly and lead to the decrease in flow stress.

6. Acknowledgement

The authors gratefully acknowledge the support of the National Key R&D Program of China (No. 2017YFB0306400), the National Natural Science Foundation of China (Grant No. 51790485), the Fundamental Research Funds for the Central Universities (No. N2007008), Nanning science and technology major special projects (No. 20191002), the Key research and development project of Shandong province (the first batch of competitive selection of major scientific and technological innovation projects, No. 2019JZZY010401), the Fundamental Research Funds for the Central Universities (No. N2007005).

7. References

[1] A. Cuniberti, A. Tolley, M. V. Castro Riglos, R. Giovachini, Influence of natural aging on the precipitation hardening of an AlMgSi alloy 2010 *Mater. Sci. Eng. A* **527** 5307–5311

[2] D. Jiang, C. Wang, Influence of microstructure on deformation behavior and fracture mode of Al-Mg-Si alloys 2003 *Mater. Sci. Eng. A* **352** 29-33

[3] M. Usta, M. E. Glicksman, R. N. Wright. The effect of heat treatment on Mg_2Si coarsening in aluminium 6105 alloy 2004 *Metall. Mater. Trans. A* **35** 435-438

[4] X. M. Qian, N. Parson, X. C. Chen, Effects of Mn addition and related Mn-containing dispersoids on the hot deformation behavior of 6082 aluminium alloys 2019 *Mater. Sci. Eng. A* **764** 138-253

[5] H. J. McQueen, S. Spigarelli, M. Kassner, E. Evagelista, Hot Deformation and Processing of Aluminium Alloys 2011 *CRC Press, Bradenton,* **FL** 87–233

[6] D. Samantaray, S. Mandal, C. Phaniraj, A. Bhaduri, Flow behavior and microstructural evolution during hot deformation of AISI Type 316 L(N) austenitic stainless steel 2011 *Mater. Sci. Eng. A* **528** 8565-8572

[7] C. J. Shi, J. Lai, X. G. Chen, Microstructural Evolution and Dynamic Softening Mechanisms of Al-Zn-Mg-Cu Alloy during Hot Compressive Deformation 2014 *Materials* **7** 244-264

[8] F. J. Humphreys, M. Hatherly, Recrystallization and Related Annealing Phenomena 2004 *second ed. Elsevier Ltd., Oxford* 169-450

[9] X. M. Qian, N. Parson, X. G. Chen, Effect of Homogenization Treatment and Microalloying with Mn on the Microstructure and Hot Workability of AA6060 Aluminium Alloys 2019 *Mater. Eng. Perform.* **28** 4531–4542

[10] X. M. Qian, N. Parson, X. G. Chen, Effect of post-homogenisation cooling rate and Mn addition on Mg_2Si precipitation and hot workability of AA6060 alloys 2020 *Can. Metall. Quart.* **59** 189-200

[11] M. Shakiba, N. Parson, X. G. Chen, Effect of Iron and Silicon Content on the Hot Compressive Deformation Behavior of Dilute Al-Fe-Si Alloys 2015 *J. Mater. Eng. Perform.* **24** 404–415

[12] R. Mann, J. Hexemer, I. W. Donaldson, Bishop D P, Hot deformation of an Al–Cu–Mg powder metallurgy alloy 2011 *Mater. Sci. Eng. A* **528** 5476-5483.

[13] M. Rajamuthamilselvan, S. Ramanathan, Hot deformation behaviour of 7075 alloy 2011 *J. Alloys Compd.* **509** 948-952

[14] C. J. Shi, J. Lai, X. G. Chen, Microstructural Evolution and Dynamic Softening Mechanisms of Al-Zn-Mg-Cu Alloy during Hot Compressive Deformation 2014 *Materials* **7** 244-264

[15] E. Cerri, E. Evangelista, H. J. McQueen. Overview of high temperature substructure development in Al-Mg alloys 1999 *High. Temp. Mat. Proc.* **18** 227–240

[16] Jazaeri, F. J. Humphreys, The transition from discontinuous to continuous recrystallization in some aluminium alloys II – annealing behavior 2004 *Acta Mater* **52** 3251–3262

[17] H. Yamagata Dynamic. recrystallization of single-crystalline aluminium during compression tests. Scripta Metall 1992 *Mater.* **27** 727–732
[18] H. Yamagata. In situ observation of dynamic recrystallization in five-nine aluminium by a transmission Laue method 1994 *Scripta Metall. Mater.* **30** 411–416
[19] X. Huang, H. Zhang, Y. Han, W. Wu, J. Chen, Hot deformation behavior of 2026 aluminium alloy during compression at elevated temperature 2010 *Mater. Sci. Eng. A* **527** 485-490
[20] S. Y. Chen, K. H. Chen, G. S. Peng, X. H. Chen, Q. H. Ceng, Effect of heat treatment on hot deformation behavior and microstructure evolution of 7085 aluminium alloy 2012 *Alloys Comp.* **537** 338-345
[21] S. Gourdet, E. V. Konopleva, H. J. McQueen, F. Montheillet, Recrystallization during hot deformation of aluminium 1996 *Mat. Sci. Forum* **217** 441-446
[22] F. Gourdet, A. Montheillet, model of continuous dynamic recrystallization 2000 *Mat. Sci. Eng. A* **283** 274-288
[23] A. G. Beer, M. R. Barnett, Microstructural development during hot working of Mg- 3Al-1Zn, Metall 2007 *Mater. Trans. A* **38A** 1856-1867
[24] E. Cerri, E. Evangelista, A. Forcellese, H. J. McQueen, Comparative hot workability of 7012 and 7075 alloys after different pretreatments 1995 *Mater. Sci. Eng. A* **197** 181-198
[25] S. Y. Chen, K. H. Chen, G. S. Peng, X. H. Chen, Q. H. Ceng, Effect of heat treatment on hot deformation behavior and microstructure evolution of 7085 aluminium alloy 2012 *Alloys Comp.* **537** 338-345
[26] H. Zhang, L. X. Li, D. Yuan, D. S. Peng, Hot deformation behavior of the new Al–Mg–Si–Cu aluminium alloy during compression at elevated temperatures 2007 *Mater. Charact.* **58** 168–173
[27] H. C. Liao, Y. N. Wu, K. X. Zhou, J. Yang, Hot deformation behavior and processing map of Al–Si–Mg alloys containing different amount of silicon based on Gleebe-3500 hot compression simulation 2015 *Mater. Des.* **65** 1091–1099
[28] H. Z. Li, Z. Li, M. Song, X. P. Liang, F. F. Guo, Hot deformation behavior and microstructural evolution of Ag-containing 2519 aluminium alloy 2010 *Mater. Des.* **31** 2171–2176
[29] Y. C. Lin, L. T. Li, Y. C. Xia, Y. Q. Jiang, Hot deformation and processing map of a typical Al–Zn–Mg–Cu alloy 2013 *Alloys Compd.* **550** 438–445
[30] M. R. Rokni, A. Zarei-Hanzaki, A. A. Roostaei, H. R. Abedi, An investigation into the hot deformation characteristics of 7075 aluminium alloy 2011 *Mater. Des.* **32** 2339–2344
[31] N. Jin, H. Zhang, Y. Han, W. Wu, J. Chen, Hot deformation behavior of 7150 aluminium alloy during compression at elevated temperature 2009 *Mater. Charact.* **60** 530–536.
[32] Shakiba M, Parson N, Chen X G, Effect of homogenization treatment and silicon content on the microstructure and hot workability of dilute Al–Fe–Si alloys 2014 *Mater. Sci. Eng. A* **619** 180–189
[33] Y. Y. Dong, C. S. Zhang, G. Q. Zhao, Y. J. Guan, Anjiang Gao, Wenchao Sun, Constitutive equation and processing maps of an Al–Mg–Si aluminium alloy: Determination and application in simulating extrusion process of complex profiles 2016 *Mater. Des.* **92** 983–997
[34] Y. N. Kwon, Y. S. Lee, J. H. Lee, Deformation behavior of Al–Mg–Si alloy at the elevated temperature 2007 *Mate. Proc. Tech.* **187** 533–536
[35] Z. Li, Z. Zhang, X. G. Chen, Effect of magnesium on dispersoid strengthening of Al–Mn–Mg–Si (3xxx) alloys 2016 *Trans. Nonferrous Met. Soc. China* **26** 2793–2799
[36] Q. L. Zhao, M. Slagsvold, B. Holmedal, Comparison of the influence of Si and Fe in 99.999% purity aluminium and in commercial-purity aluminium 2012 *Scripta Mater.* **67** 217-220
[37] J. Zhang, F. S. Pan, R. L. Zuo, C. G. Bai, The low temperature precipitation in commercial-purity aluminium sheets for foils 2008 *J.Mater. Process. Tech.* **206** 382-387
[38] M. Shakiba, N. Parson, X. G. Chen, Hot deformation behavior and rate-controlling mechanism in dilute Al–Fe–Si alloys with minor additions of Mn and Cu 2015 *Mater. Sci. Eng. A* **636** 572–581
[39] G. Neumann, C. Tuijn, Self-Diffusion and Impurity Diffusion in Pure Metals: Handbook of Experimental Data 2009 *Elsevier Ltd.* 126-150

[40] G. Sha, Y. B. Wang, X. Z. Liao, Z. C. Duan, S. P. Ringer, T. G. Langdon, Influence of equal-channel angular pressing on precipitation in all A1-Zn-Mg-Cu alloy 2009 *Acta Mater.* **57** 123-132

[41] M. Legros, G. Dehm, E. Arzt, T. J. Balk, Observation of giant diffusivity along dislocation cores 2008 *Scienc.* **319** 1646-1649

[42] Y. Birol, Effect of cooling rate on precipitation during homogenization cooling in an excess silicon AlMgSi alloy 2012 *Mater. Charact.* 37-42

[43] J. Langerweger. Effect of metallurgical factors on productivity in the extrusion of aluminium magnesium silicon (AlMgSi) alloys 1982 *Aluminium* **58** 107-109.

IOP Conf. Series: Materials Science and Engineering 1133 (2021) 012008 doi:10.1088/1757-899X/1133/1/012008

Experimental research on four-point-bending loading of composite fuselage frame

Li Xiang[1a], **Li Xingfu**[2]
[1]China Southern Airlines Co, Ltd. Shenzhen Branch, Shenzhen 518000, China;
[2]AVIC Xi 'an Aircraft Industry Group Co. Ltd, Xi'an 710089, China
[a]xiangleenwpu@aliyun.com

Abstract. A four-point-bending loading test device was designed for the pure bending loading of the composite fuselage curved frame. The reinforcing subplates, reinforcing frames, wedge-shaped blocks and other parts were installed in the load segment of test pieces to enhance the local stiffness, strength and torsional stiffness. Four-point-bending loading tests and nonlinear finite element calculation of C-frame and Z-frame pieces under the condition of positive bending was carried out. The calculated buckling mode was in good agreement with the test result, which provides a certain theoretical basis for the instability mode analysis of curved frames. The test and analysis results show that under the condition of positive bending and the middle section of C-frame has lateral instability and the end inner edge of the frame has the local buckling and damage.While the middle section of inner edge of the Z-frame has only local buckling and damage. The test results of two kinds of frame show that the stability and structural strength of Z-frame are better than that of C-frame.

1. Introduction

Composite materials have been used more and more widely in aircraft structure because of their excellent performance. Among which the most typical composite structure is the fuselage panel composed of skin, girder and frame, etc. The girder mainly sustain the action of axial tension and compression in the fuselage structure while the role of the frame is to increase the lateral stability of the fuselage. In other words, when the panel structure composed of skin and girder is partially buckling, the circumferential connected frame can still support the girder and avoid the overall instability of the fuselage panel. The fuselage is equivalent to a closed annular barrel segment and the aerodynamic load borne by the skin is transmitted to the frame through the girder and making the frame sustain a large bending load. Therefore, the stability of the frame is very important for the whole fuselage. It is of great significance for the fuselage design to study the circumferential stability of the frame and assess the instability mode of the frame [1].

At present, experimental studies in domestic on fuselage structure mainly focus on axial tension-compression performance and circumferential charging performance [2]-[3]. Most experimental studies on circumferential loading performance of fuselage frame are carried out by single-frame bending loading method [4]-[7]. Gao Binhua [5] used the three-point bending loading method to carry out dynamic loading on the J-type single frame of braided composite material. The section of the J-type frame was bent and twisted and the load contact surface at the edge of the frame was broken first. Current research mainly focuses on the bending loading of single or straight frames and the assessment objective is simple. The experimental support conditions are greatly different from the real fuselage structure. So it is difficult to verify the real instability mode of fuselage frame with circumferential

IOP Conf. Series: Materials Science and Engineering 1133 (2021) 012008 doi:10.1088/1757-899X/1133/1/012008

load. Moreover, the support stiffness of the frame is weak and problems such as roll-over, torsion and premature failure are easy to occur when the concentrated force is applied.

The actual boundary condition of the frame is elastic support of skin and girder on one side and the instability mode of the frame is quite different from that of the single frame loading. In order to study the circumferential stability of the frame actually, we design a frame structure composed of skin, stringer, shear angle sheet and curved frame, using a four-point bending load and designing reasonable way of connection and support to avoid frame side lean, torsion and early damage during the process of loading and reaching the purpose of assessment of the frame of instability and failure. At the same time, the experimental results were analyzed with finite element calculation and the instability modes and failure modes of C-frame and Z-frame were obtained respectively and the optimal configuration was given through comparison, which is of great significance to the structure design of composite fuselage.

2. Test pieces

2. 1. Test pieces configuration

The test pieces include frame, shear angle sheet, girder and skin, etc. The test section is 3 times girders length in the middle of frame and the load segment is 1.5 times girders length at both ends. The test section is same with the actual fuselage structure. The clamping fixture is considered in the design of the load segment. The original girder is removed and a layer of subplate is laid on the inner surface of the skin. The skin radius of the test pieces is 2960mm and the interception span is 28. 455°and the horizontal length is about 1455mm and the width is 620mm. The basic appearance is shown in Figure 1. The test pieces contains C-frame and Z-frame cross-section configurations. In which the skin, girder and shear angle sheet configurations are the same. As shown in Figure 2. The inner edge extension directions of the frame of the two cross-section configurations are opposite.

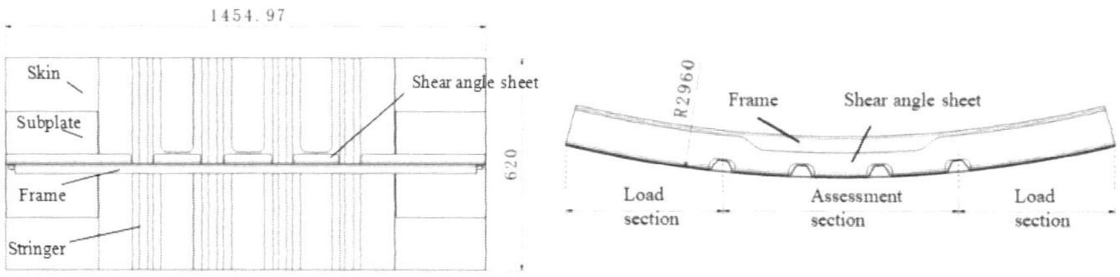

(a) Front View of Test Pieces (b) Side View of Test Pieces

Figure 1. Sketch of Fuselage Curved Panel

(a) C-Frame Section Configuration (b) Z-Frame Section Configuration

Figure 2. Frame Section Configuration

2. 2. Material of test pieces

Every part of the test pieces are made of M21E/IMA. Material properties is E_{11}=154GPa, E_{22}=8.5GPa, E_{33}=8.5GPa, G_{12}=4.2GPa, ν_{12}=0.35. The thickness of the single layer is 0.184mm. Among them, the layer order of skin is [45/-45/-45/90/45/0]s. The layer order of cap stringer is [45/0/0/-45/90/-45/0/0/45]. The layer order of shear angle sheet is [45/-45/0/90/-45/45/0]s. The layer order of frame is [45/-45/0/90/45/-45/90/0/-45/45].

3. Test and calculation method

3. 1. Load method

In this test, four-point bending loading is adopted for pure bending loading and the schematic diagram of the loading scheme is shown in Figure 3. The two ends of the test pieces are placed on two support seats respectively and the two loading seats exert concentrated loads on both sides of the assessment section respectively. Wedge-shaped blocks are installed at the support and loading places to convert the curved surface into a flat so that the supporting force and the loading force are parallel so as to form a pair of force couplings of equal size and opposite direction and thus form pure bending load. In this test, the positive bending load is applied to the frame of the fuselage. So the support seat is at both ends of the test pieces and the loading seat is in the middle. Among them, the horizontal distance between the support seat and the loading seat is L. If the load applied by the loading seat is F/2, then the bending moment suffered by the assessment section is $M=1/2F\cdot L$.

3. 2. Support strengthening method

As the local stiffness and strength of the frame and skin are weak, it is easy to be unstable or destroyed advanced under the action of concentrated load. So it is necessary to strengthen the stiffness and strength of the whole load segment. In this test, a variety of reinforcing frames and reinforcing subplates are connected in the load segment to form a set of strengthening components. The section combination is shown in Figure 4. In Figure 4, the left and right side of shear angle sheet connect C-reinforcing frames and L-type reinforcing frame 2. And in order to increase the support stiffness, L-type reinforcing frame 1 is placed on the right side of connections. In order to increase local stiffness and strength of web and inner of frame, the reinforcing blocks are connected at the top of the frame to increase the strength of contact surface. The support subplate connects at the bottom of skin to increase the local stiffness and strength of support surface skin. All the connecting surfaces adopt the form of surface matching and mechanical connection to ensure sufficient connection strength.
The reinforcing frames and reinforcing subplates connected in the load segment are not only increasing the supporting stiffness but also effectively increasing the torsional stiffness and stiffness symmetry of the section. In addition, the transverse support surface of the loading surface and support surface is large, which effectively increases the transverse support stiffness of the section. The loading seat and support seat are fixed on the test machine, so the transverse boundary support of the loading surface and support surface is similar to the fixed support, which effectively limits the torsion or deformation of the section (As shown in Figure 5.).

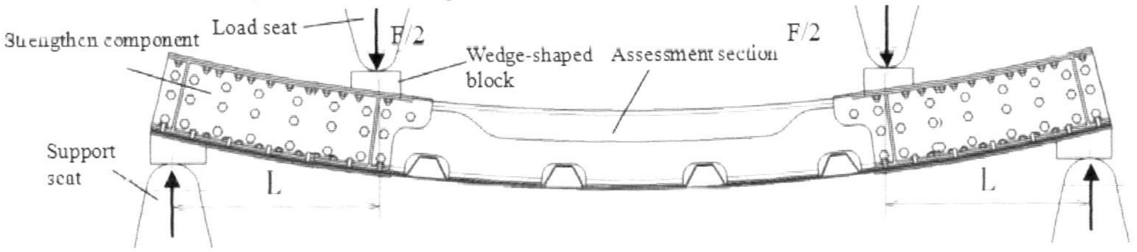

Figure 3. Four-point bending loading diagram of fuselage curved panel

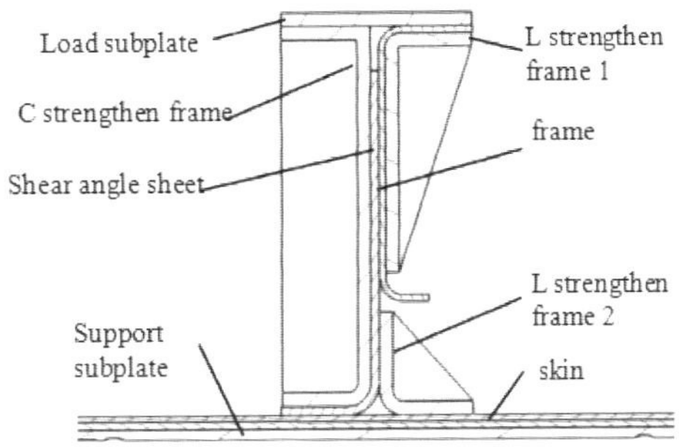

Figure 4. Section Sketch of Load Segment

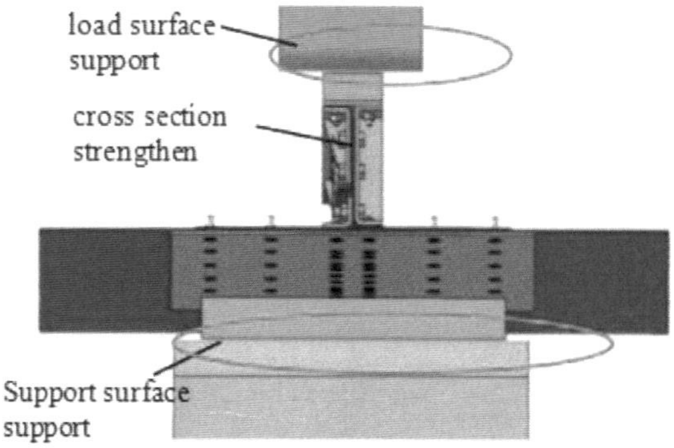

Figure 5. Support Sketch of Load Segment

3. 3. Test measurement and implementation

Four-point bending test of fuselage frame was carried out on instron500kn testing machine and the test photos are shown in Figure 6. Upper and lower chuck of testing machine fixed support platform and install the load seat close to inside and support seat close to outside on the upper and lower support platform respectively and test pieces placed on the support seat. The support platform and two load seats transfer load from testing machine to load points of assessment segment of test pieces on both sides to form bending load. The horizontal distance between the load point and the support point in the test was L=309mm. According to the description in section 2.1, if the load value of the testing machine is F (unit KN), then the actual bending moment of the testing section of the test pieces M=154. 5F (KN·mm). The load method adopts displacement loading and the load rate is 1mm/min and the accuracy is level 0. 5.

Figure 6. Four-point bending of fuselage frame

In order to measure the strain response of the test pieces during the loading process, a strain gauge is pasted on the concerned part as shown in Figure 7. Among them, 20-26 pieces are pasted on the inner surface of the inner edge of the frame. 27-31 pieces are pasted on the side of the web of the frame. 40-42 pieces are pasted on the side of the shear angle sheet. And 22-24 pieces are pasted on the side of the middle part of the inner edge of the frame. All strain gauges are pasted back to back along circumferential of the frame. The number of strain gauges on the back is +100. For example, the No. 20 piece is located on the inner surface of the inner edge of the frame And the No. 120 symmetrical pieces on the back is located on the outer surface of the inner edge of the frame. St-16 strain acquisition system is adopted for strain measurement with range of ±20000με and measure accuracy of 0. 5% FS.

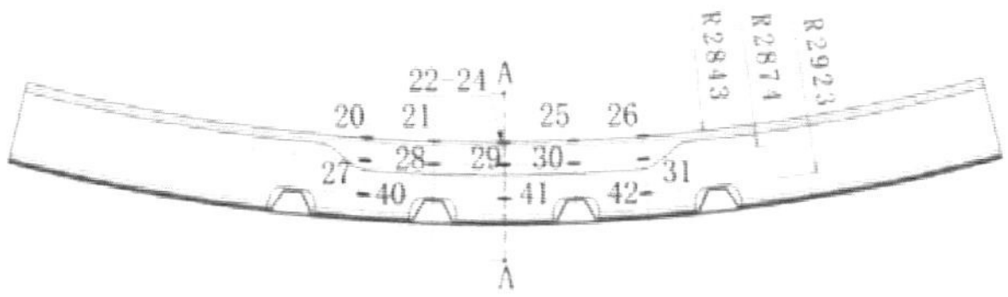

(a) Strain Gauge Arrangement

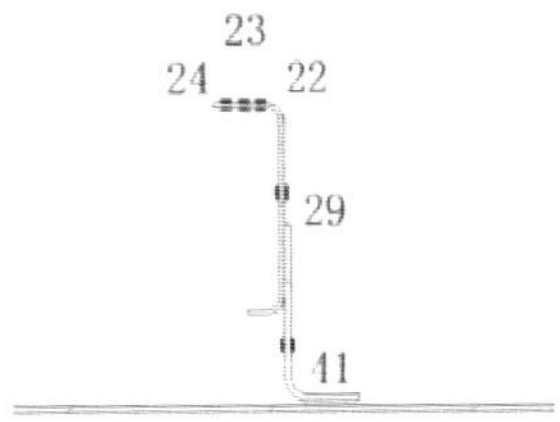

(b) A-A Cross-Section Strain Gauge Arrangement

Figure 7. Strain Gauge Arrangement on Specimen

3. 4. Finite element analysis method

Using ABAQUS software, the finite element calculation based on the actual connection relationship and boundary conditions and load conditions of the test is carried out to assist the analysis of the instability mode of the test pieces. In the finite element model, the test pieces and the reinforcing frames and the reinforcing subplates all adopt shell element. The wedge-shaped block and the supporting seat adopt solid element. The bolt connection adopt 3D beam element and node coupling mode while the support surface and the load surface adopt face-contact. When the boundary conditions were set two support seats were fixed and two load seats were loaded with vertical downward displacement. Symmetrical boundary constraints were set on the circumferential central of the test piece. The finite element model was shown in Figure 8. The finite element analysis verify the instability mode of the test pieces only. Thus only geometric nonlinear calculation is adopted and the calculation of material damage is not considered.

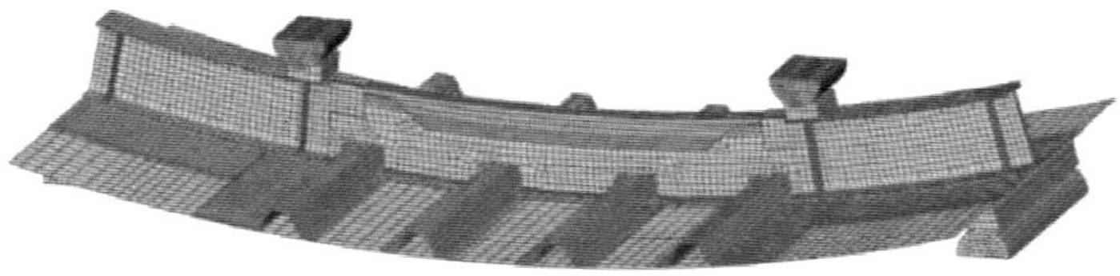

Figure 8. The Finite Element Model of Fuselage Curved Frame

4. Results and analysis

4. 1. C-Frame test results analysis

When the fuselage frame is loaded with positive bending, the frame mainly sustain compression load, which will cause instability. The load-strain curve of the instability position drawn according to the positive bending loading test results of C-frame test pieces is shown in Figure 9. The abscess of the figure is the load F directly applied by the testing machine.

From Figure 9 (a), it can be found that c-frame test assessment within the inner edge of the middle around the lateral surface of three strain gauges (22, 23, 24) of the load and strain curve showed a trend of bending upwards of nonlinear. Based on literature [7] that the lateral buckling happened, it namely the lateral deformation instability along the inner edge of frame and not normal deformation instability. The lowest point of the intermediate strain gauge is defined as the lateral instability point, then the lateral instability load of the frame is 17. 8kn and the instability strain is -1875με. It can be seen from Figure. 9 (b) that the load-strain curve of back-to-back strain gauges (pieces 20 and 120) at the inner edge of the frame at the end of the assessment section generated divarication and inflexion point at 19. 8kn and the inflexion point is defined as the local buckling point, and the average value of back-to-back strain gauges is defined as the local buckling strain. The local buckling load at the inner edge of the end frame is 19. 8kn and the buckling strain is -3120με. When the test was loaded up to 24kn, failure occurred at the local buckling position (no. 20 and no. 120 pieces) at the inner edge of the end of frame and the failure mode was shown in Fig 10. In summary, when the c-frame test pieces were positively flexed under loading, lateral instability occurred in the middle section of the test section first, and led to local buckling at the inner edge of the end frame. With the increase of load, fracture failure occurred at the local buckling part of the end.

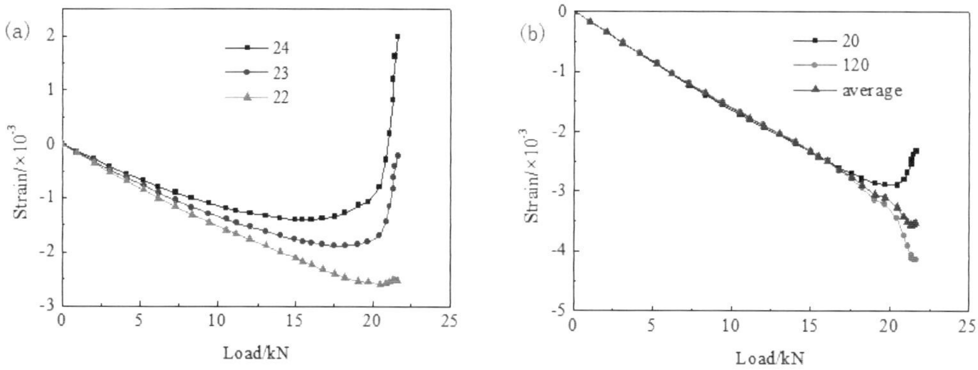

Figure 9. Test Instability Load-Strain Curves of C-Frame: (a) Lateral Instability; (b) Local Instability

The instability mode of c-frame test pieces under positive bending loading obtained by geometric nonlinear finite element calculation is shown in Figure 11. From the figure, it can be intuitively found that lateral instability deformation occurred in the middle section of the assessment section and local buckling deformation occurred at the inner edge of the end. The instability load-strain curves of the middle segment of the inner edge and the end of the c-frame calculated by the finite element are shown in Figure 12 and the strain collection site in the figure is completely the same with the test pieces. According to the analysis method of the test results, the lateral buckling load and buckling strain of the frame calculated by the finite element method are 20. 5kn, -1950με. The local buckling load at the inner edge of the end frame is 22kn and the buckling strain is -3280με, which are close to the experimental results and the maximum error is less than 15%. It can be seen that the instability mode and results of finite element calculation are in good agreement with the test and the instability mode of the structure can be analyzed more intuitively.

Figure 10. Test Failure Mode of C-Frame

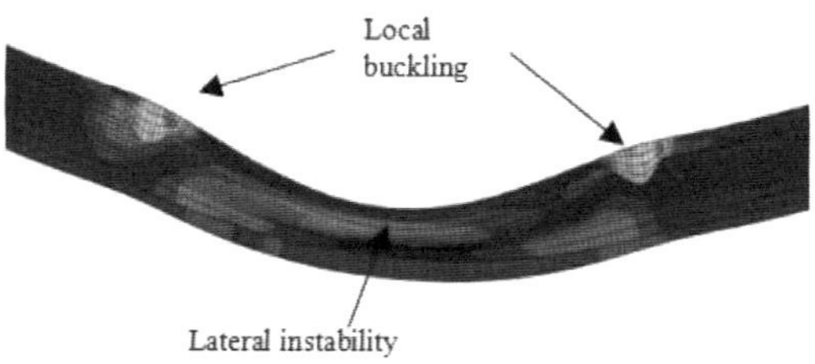

Figure 11. FEA Instability Mode of C-Frame

4. 2. Z frame test results analysis

The mid-section load-strain curve of Z-frame test pieces under positive bending loading obtained from the test is shown in Figure 13. It can be seen from Figure 13 (a) that the load-strain curves of the three lateral strain gauges (pieces 22, 23 and 24) at the inner edge of the middle section of the Z-frame do not show obvious non-linearity and indicate that lateral instability does not occur at the middle section of the inner edge. However, it can be seen from Figure 13(b) that the load-strain curves of back-to-back strain gauges (plates 24 and 124) at the inner edge of the middle section of the Z-frame show an obvious bifurcating trend and indicate that local buckling occurs here. The buckling load is 23. 5kN and the buckling strain is -3500με (taking the average value of the back-to-back 24 and 124 plates). When the test was loaded up to 28kN, the local buckling part of the inner edge of the middle section of the Z-frame was damaged and the failure mode was shown in Figure 14. It can be seen that initial buckling occurs at the local inner edge of the frame in the middle section of the test section when the Z-frame is loaded with positive bending and rupture occurs at the local buckling part of the inner edge of the frame in the middle section with the load increase.

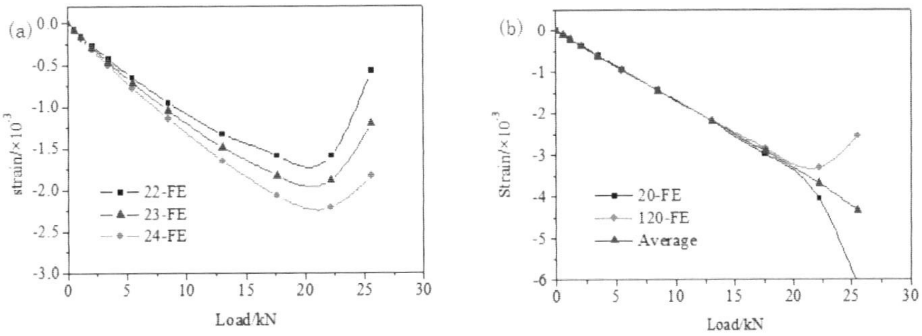

Figure 12. FEA Instability Load-Strain Curves of C-Frame: (a) Lateral Instability; (b) Local Instability

The instability mode of Z-frame under positive bending loading obtained by finite element calculation is shown in Figure 15. It can be found from the figure that corrugated local buckling occurs along the normal direction at the inner edge of the frame in the middle section of the assessment section. It can be seen from Fig 16 that the variation trend of load-strain curve of Z-frame instability calculated by finite element method is in good agreement with the test results. According to the analysis method of test results, the buckling load calculated by finite element method is 22. 5kN and the buckling strain is -3806μεand the maximum error between the finite element calculation and the test is less than 10%.

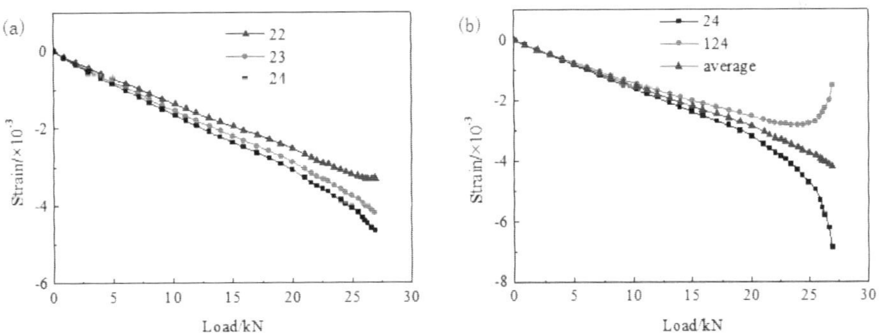

Figure 13. Test Load-Strain Curves in The Middle of Z-Frame: (a)Lateral; (b) Instability

Figure 14. Test Failure Mode of Z-Frame

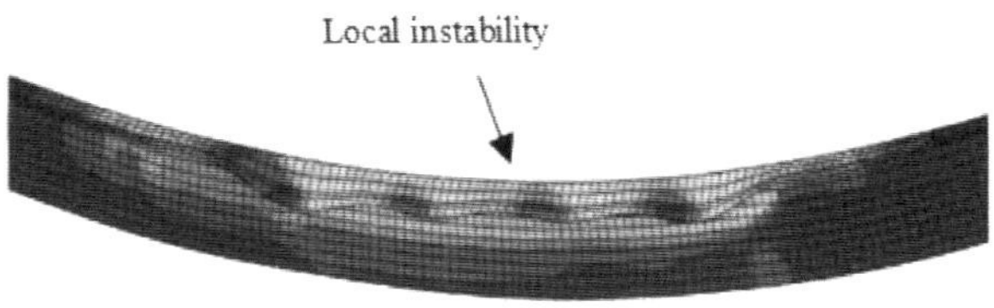

Figure 15. FEA Instability Mode of Z-Frame

4. 3. Comparison of frame configurations

It can be found through tests and finite element analysis that when the frame of fuselage is under positive bending loading that the C-frame is prone to lateral bending and lateral instability due to the asymmetry of the section when it is subjected to the compression load brought by bending, which leads to the stress concentration at the inner edge of the end and fracture. The section of Z-frame is relatively symmetrical and it is not easy to bend laterally when subjected to the compression load caused by bending. Local buckling and failure only occur at the inner edge of the mid-section frame, which belongs to the ideal instability mode and failure mode. According to the test data analyzed in Section 3. 1 and Section 3. 2, it can be found that the initial buckling load, buckling strain and failure load of Z-frame structure are 32%, 86. 7% and 16. 7% higher than that of C-frame structure under positive bending condition respectively. Therefore, from the perspective of strength and stability of composite fuselage structure, Z-frame configuration of fuselage structure is better than C-frame configuration.

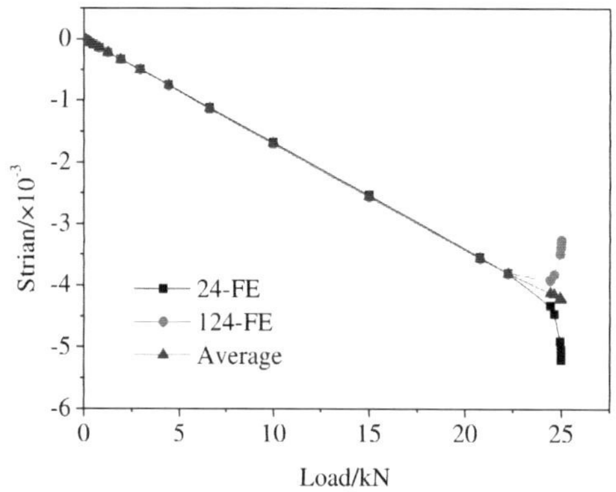

Figure 16. FEA instability load-strain curves of Z-frame

5. Conclusion

1)The four-point bending test technology of fuselage frame studied in this paper solves the problems of side lean, torsion, and advanced instability and damage in the loading section of frame and achieves the purpose of the instability and damage in the assessment section;

2) Under positive bending loading of C-frame, lateral instability first occurs in the middle section of the assessment section and leads to local buckling and fracture failure at the inner edge of the end. The initial buckling occurred at the inner edge of the frame in the middle section of the test section and was damaged at the local buckling position of the inner edge of the middle section when the Z-frame was loaded with positive bending

3) Under positive bending condition, the initial buckling load, buckling strain and failure load of Z-frame are 32%, 86. 7% and 16. 7% higher than those of C-frame test pieces. In terms of structural strength and stability, Z-frame configuration is superior to C-frame configuration.

6. References

[1] Maher B., Tarek L., Mohamed S. Mechanical Response of a Hexagonal Grid Stiffened Design of a Pressurized Cylindrical Shell-Application to Aircraft Fuselage[J]. Thin-Walled Structures, 2018(127): 40-50.

[2] ZE L Y, PU X. Crashworthiness Study of Composite Fuselage Section[J]. Key Engineering Materials, 2016(725): 94-98.

[3] CHEN An, WEI Yu-long, LIAO Jiang-hai. Damage Tolerance Test of Stiffened Fuselage Panel Under Complex Load[J]. Acta Aeronautica ET Astronautica Sinica, 2017, 38(1): 420093 (in Chinese)

[4] XIN Ya-jun, XIAO Bo, CHENG Shu-liang. Experimental Study of Four-point Bending Performance of Aluminum Foam Sandwich Beam[J]. Journal of Experimental Mechanics, 2016, 31(5): 593-599(in Chinese)

[5] XIN Ya-jun, XIAO Bo, CHENG Shu-liang. Experimental Study of Four-point Bending Performance of Aluminum Foam Sandwich Beam[J]. Journal of Experimental Mechanics, 2016, 31(5): 593-599(in Chinese)
[6] GAO Bin-hua, REN Yi-ru. Impact Dynamic Characteristics of Braided Composite Fuselage Frame[J]. Acat Material Composite Sinica, 2017, 34(8):1780-1787 (in Chinese)
[7] Lei Zhang, Gen-shuTong. Lateral Buckling of Simply Supported C-Section and Z-Section Purlins with Top Flange Horizontally Restrained. Thin-Walled Structures, 2016 (99): 155–167.

AUTHOR INDEX